# EINSTEIN'S *OTHER* THEORY

# Einstein's *Other* Theory

## THE PLANCK-BOSE-EINSTEIN THEORY OF HEAT CAPACITY

*Donald W. Rogers*

PRINCETON UNIVERSITY PRESS   PRINCETON AND OXFORD

Library of Congress Cataloging-in-Publication Data

Rogers, Donald, 1932–
    Einstein's other theory: the Planck-Bose-Einstein theory of heat capacity/Donald W. Rogers.
        p. cm.
    Includes bibliographical references and index.
    ISBN 0-691-11826-4 (acid-free paper)
    1. Blackbody radiation. 2. Specific heat. 3. Bose-Einstein condensation. I. Title:
Planck-Bose-Einstein theory of heat capacity. II. Title.

    QC484.R64 2005
    530.4′2—dc22
                        2004042068

British Library Cataloging-in-Publication Data is available

This book has been composed in Sabon

Printed on acid-free paper. ∞

www.pup.princeton.edu

Printed in the United States of America

10 9 8 7 6 5 4 3 2 1

THERE WERE GIANTS IN THE EARTH
IN THOSE DAYS . . . .

—Genesis 6:4

# Contents

*Preface*                                                                    xiii

**One. History**
1.1   Failure of the Dulong-Petit Law                                          1
1.2   Crystals: Einstein's View                                                2
      PROBLEMS                                                                  2

**Two. Background**
2.1   Classical Harmonic Motion                                                4
2.2   Wave Equations: The Vibrating String                                     7
2.3   Wave Motion                                                              9
2.4   Solving the Wave Equation: Separation of Variables                      11
2.5   The Time-Independent Wave Equation $X(x)$                                12
2.6   The Time-Dependent Wave Equation $T(t)$                                  15
2.7   Superpositions                                                          16
2.8   A Two-Dimensional Wave Equation                                         17
2.9   The Time-Independent Wave Functions $X(x_1)$ and $X(x_2)$               18
2.10  A Three-Dimensional Wave Equation                                       20
2.11  The Boltzmann Distribution                                             21
2.12  Degrees of Freedom                                                     23
2.13  Kinetic Energy per Degree of Freedom                                   24
2.14  Boltzmann's Constant                                                   26
2.15  The Translational Energy                                               27
2.16  The Energy of a Vibrational State Is $k_B T$                           27
2.17  Trouble Brewing                                                        28
      PROBLEMS                                                               29

**Three. Experimental Background**
3.1   Thermal Radiation in a Chamber or Cavity                               30
3.2   Kirchhoff's Law: Absorptivity                                          33
3.3   The Intensity of Radiation                                             34
3.4   The Stefan-Boltzmann Law: Emissivity                                   36
3.5   Stefan's Law                                                           36
3.6   The Blackbody Radiation Spectrum                                       41
3.7   Measurement of the Blackbody Spectrum                                  43
3.8   Astrophysical Data from the COBE Satellite                             46
      PROBLEMS                                                               48

**Four. The Planck Equation**
4.1   The Paschen-Wien Law                                            49
4.2   Fitting the Curve                                               51
4.3   The Number Density of Oscillatory Modes                        56
4.4   The Rayleigh-Jeans Equation                                    60
4.5   The Planck Equation                                            62
4.6   Immediate Deductions from Planck's Law                         67
      PROBLEMS                                                       68

**Five. The Einstein's Equation**
5.1   The Einstein Model                                             70
5.2   Einstein's First Derivation: The Heat Capacity of Diamond      71
5.3   The Einstein Temperature                                       73
5.4   Difficulties with the Einstein Theory                          75
      PROBLEMS                                                       76

**Six. The Debye Equation**
6.1   The Debye Model                                                77
6.2   The Debye Equation                                             79
6.3   The Debye Temperature                                          82
6.4   The Integral $D$                                               84
6.5   Very-Low-Temperature Behavior of the Debye Equation           84
6.6   The Speed of Sound in Solids                                   86
6.7   The Debye Third-Power Law                                      90
6.8   Third-Law Entropies                                            90
      PROBLEMS                                                       92

**Seven. Quantum Statistics**
7.1   The Photoelectric Effect                                       94
7.2   The Photon Gas                                                 96
7.3   Bose's Letter to Einstein                                      97
7.4   The Quantum Harmonic Oscillator                                98
7.5   The Total Vibrational Energy                                   101
7.6   Heat Capacity                                                  101
7.7   Bosons and Fermions                                            103
7.8   Permutations and Combinations                                 104
7.9   Configurations                                                 105
7.10  Stirling's Approximation                                       107
7.11  Constraints                                                    107
7.12  The Classical Boltzmann Distribution                           108
7.13  Fine Structure                                                 110
7.14  The Classical Case: A More General Derivation                  112
7.15  Fermi-Dirac Counting                                           113

7.16 The Fermi-Dirac Distribution Function                               116
7.17 Bose-Einstein Counting                                             117
7.18 The Bose-Einstein Statistical Weights $W_B$                         119
7.19 The Bose-Einstein Distribution Function                            120
7.20 Summary Equations                                                 121
7.21 An Alternative Derivation for Fermions and Bosons                  121
7.22 Fermions (Again)                                                   123
7.23 Bosons (Again)                                                     123
7.24 Reduction to the Classical Case                                    125
7.25 The Entropy                                                        126
7.26 A Note from Classical Thermodynamics:
    The Fundamental Equation                        129
    PROBLEMS                                         129

**Eight. Consequences of the Fermi-Dirac Distribution**
8.1  The Electron Gas                                                   132
8.2  The Fermi Sea                                                      132
8.3  The Fermi Distribution                                            135
8.4  The Electronic Contribution to Solid-State Heat Capacity           136
8.5  The Ground State of a Fermi Gas                                    137
8.6  The Number of Orbitals in the Ground State                         139
8.7  The Total Energy of Electrons in the Ground State                  140
8.8  The Density of States                                              141
8.9  The Energy of an Electron Gas                                      144
8.10 The Low-Temperature Heat Capacity of an Electron Gas               145
8.11 The Debye-Sommerfeld Equation                                      146
    PROBLEMS                                         148

**Nine. Consequences of the Bose-Einstein Distribution**
9.1  Of Waves and Particles                                            149
9.2  Bose: The Density of Photon Modes                                 150
9.3  Why Is $\mu = 0$ for Photons?                                     152
9.4  Phonons                                                           154
9.5  The Influence of Symmetry Numbers on
    Rotational Spectroscopy                         155
9.6  The Vibrational Partition Function $q_{\text{vib}}$               157
9.7  The Rotational Partition Function $q_{\text{rot}}$                158
9.8  Symmetry Numbers                                                  159
9.9  Bosons, Fermions, and Triplets                                    161
9.10 The Einstein Coefficients                                         162
9.11 Lasers                                                            164
9.12 The Bose-Einstein Condensation                                    165

9.13 The Bose-Einstein Condensation of Metal Vapor
        (Nobel Prize, 2001)                              165
9.14 Ballistic Expansion                                167
9.15 Macroscopic Quantum Effects                        168
9.16 Superfluidity                                      168
9.17 Order Parameters                                   170
9.18 Superconductivity                                  171
9.19 Stopped Light                                      172
9.20 Vortices                                           173
        **PROBLEMS**                                    175

*Bibliography*                                          177

*Index*                                                 179

# Preface

Say "Einstein" to almost anyone in the literate world, and the response will undoubtedly be "relativity." Relativity has captured the imagination of thoughtful people both inside and outside the scientific community as no mathematical theory has since the time of Isaac Newton. We struggled, not so much to grasp, but to *believe* the weird predictions of relativity theory—shrinking space ships and bending light beams—and we watched with fascination as Einstein's predictions were relentlessly and unequivocally verified by a hundred experiments and astronomical observations.

What of Einstein's *other* great theory, that of the Bose-Einstein condensate? The last decade of the twentieth century has produced results that rival in weirdness the space-time contractions of relativity theory. This book will trace the history of radiation and heat capacity theory from the mid-nineteenth century to the present. We will start with attempts to understand heat and light radiation, proceed through the theory of the heat capacity of solids, and arrive at the theory of superconductivity, the astonishing property of some liquids to spontaneously crawl up and out of any container, and the ability of some gases to cause light to stop for a moment's rest from its inexorable forward flight.

Are these theories and properties really weird? Indeed, *can* the universe be weird? In fact, the natural universe is as we find it, neither weird nor nonweird. Nature is nature. If, when looking upon nature we think we see weirdness, what we really see is a reflection of our own limited imagination. Let us put those limitations aside as best we can, and take an unbiased look at the Planck-Einstein theory of blackbody radiation, the Einstein theory of solid-state heat capacities, Bose-Einstein statistics, and Bose-Einstein condensation.

I wish to thank Joe Wisnovsky, formerly of Princeton University Press, for his unfailing belief that this book would be a worthy enterprize.

Greenwich Village
New York

# EINSTEIN'S *OTHER* THEORY

# One

## History

IN 1819 DULONG AND PETIT enunciated a principle, which now bears their names, that the atomic weight of a solid element times its specific heat is a constant. In modern units,

$$\text{at wt.} \times \text{sp. heat} \cong 6\,\text{cal}\,\text{K}^{-1}\,\text{mol}^{-1} \cong 25\,\text{J}\,\text{K}^{-1}\,\text{mol}^{-1}$$

where J is the joule and K is the kelvin. The modern term for (at wt. × sp. heat) is the *molar heat capacity*, designated $C_V$ when measured at constant volume. (In the nineteenth century there was some dispute over whether the heat capacity should be measured at constant pressure or constant volume, but it soon became clear that $C_V$ is more appropriate for our purposes.)

### 1.1 Failure of the Dulong-Petit Law

Historically, the law of Dulong and Petit settled several disputes about the atomic weights of solids, but we are principally interested in cases in which it fails. Boltzmann argued that the specific heat of a system can be rationalized on the basis of $\frac{1}{2}\kappa T$ of energy per degree of freedom of molecular motion where $\kappa$ is a suitable constant (see below) and $T$ is the temperature. Unfortunately, by the time this ingenious argument had been put forth, numerous violations of the law were known and the whole argument was discredited. Boltzmann's explanation was substantially correct but something radical had to be done to modify it.

Einstein noticed that the law of Dulong and Petit fails badly for diamond. Subsequent low-temperature studies showed that it always fails, provided the temperature is low enough. Whenever the law fails for simple crystals, the observed Dulong and Petit constant is smaller than it "ought to be." Later cryoscopic studies showed that the Dulong and Petit constant is always "too small" at some temperature and that it approaches zero near 0 K, a temperature region unattainable at the time of Einstein's original work. Einstein set out only to remedy problems in predicting the heat capacity of diamond but in so doing he developed a general theory of the variation of $C_V$ with $T$ for all solids at all temperatures, even down to 0 K.

## 1.2 Crystals: Einstein's View

Einstein contemplated a model of harmonic oscillators free to move in three dimensions tethered to regularly spaced lattice points by isotropic restraining forces. By isotropic, we mean that there are no differences in the forces in the $x$, $y$, and $z$ directions.

By the early twentieth century, it was known that the energy of a *classical* mechanical system is $\frac{1}{2}\kappa T$ per degree of freedom per particle or $\frac{1}{2}RT$ per degree of freedom per mole, where $R$ is the ideal gas constant, which has a modern value of $8.314\,\mathrm{J\,K^{-1}\,mol^{-1}}$. The three-dimensional harmonic oscillator has three degrees of freedom contributing to its kinetic energy and three degrees of freedom contributing to its potential energy, leading to $\frac{6}{2}RT = 3RT$ of energy. From classical thermodynamics,

$$C_V = \left(\frac{\partial U}{\partial T}\right)_V \tag{1.2.1}$$

where $U$ is the molar energy; hence

$$C_V = 3R = 3(8.31) = 24.9\,\mathrm{J\,K^{-1}\,mol^{-1}}.$$

So it is that the law of Dulong and Petit is consistent with classical mechanics. The question is: Why does it fail for diamond? This Einstein deduced in 1906 on the basis of the recent (1900) quantum theory of Max Planck. It soon became evident that his deduction is valid for very many solids at low temperatures.

If the harmonic oscillators in a simple crystal have discrete levels of energy as Planck proposed, then they should all be at the lowest level at $T = 0\,\mathrm{K}$. Very near that temperature, at $T \cong 0\,\mathrm{K}$, an infinitesimal temperature rise would not be sufficient to promote any of the atoms in the crystal to the next higher (first excited) quantum state. No promotion, no energy absorption. If there is no energy absorption for an infinitesimal temperature rise,

$$dU = C_V\,dT = 0 \tag{1.2.2}$$

and, since $dT \neq 0$, $C_V$ must be zero.

Now we have the two extremes of heat capacity for a simple crystal. At very low temperatures $C_V \cong 0$ and at high temperatures $C_V \cong 24.9\,\mathrm{J\,K^{-1}\,mol^{-1}}$. The question is what happens in between.

## PROBLEMS

**1.1.** How much heat energy (in joules) does it take to heat a kilogram of copper from 20.0 to 40.0 K?

**1.2.** How much heat energy in joules does it take to heat a kilogram of water from 20.0 to 40.0 K? Recall (or look up) the historic definition of the calorie.

**1.3.** Do the answers to problems 1.1 and 1.2 give scientific support to the folk saying, "A watched pot never boils"?

# Two

## Background

THE SIMPLE MECHANICAL SYSTEM of the classical harmonic oscillator underlies important areas of modern physical theory. This chapter develops the basic model in one, two, and three dimensions. The concept of degeneracy arises in the two-dimensional oscillation of a square plate or diaphragm. Three-dimensional harmonic oscillation relates to oscillatory modes in the Rayleigh-Jeans equation (section 4.6). Vibration of a macroscopic three-dimensional crystal is treated by Debye's theory in chapter 6. Harmonic oscillator theory is important when it succeeds and also when it fails, as we shall see in the motivation to find a theory of radiation that we now call the quantum theory, described at the end of this chapter.

### 2.1 Classical Harmonic Motion

Motion of a mass $m$ under the influence of a Hooke's law force $f$ acting on $m$ such that it moves back and forth along a straight line across an equilibrium point is *simple harmonic motion* (SHM):

$$f = -kx \qquad (2.1.1)$$

(see figure 2.1.1) In a one-dimensional $x$ space, the force is directly proportional and opposite to the displacement $x$ of the mass away from the equilibrium position, which we define as $x = 0$. At $x = 0$, $f = 0$. The proportionality constant $k$ is called a Hooke's law force constant. The sign on the right is negative because the force is opposite to the displacement, that is, it is a *restoring force*.

According to Newton's second law,

$$f = ma, \qquad (2.1.2)$$

where the acceleration $a$ is $d^2x/dt^2$ for motion along the single space coordinate $x$. Equating Hooke's law and Newton's law,

$$m\frac{d^2x}{dt^2} = -kx,$$
$$\frac{d^2x}{dt^2} = -\frac{k}{m}x, \qquad (2.1.3)$$

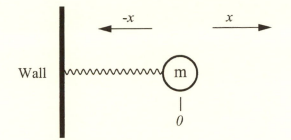

Figure 2.1.1. Harmonic oscillator of one mass.

which is the Newton-Hooke equation of motion for a harmonic oscillator of a single mass. Equation 2.1.3 has, as one of its solutions,

$$x = Ae^{i\omega t} \tag{2.1.4}$$

where $\omega = \sqrt{k/m}$ and $i = \sqrt{-1}$. Sine and cosine solutions like $x = A \sin \omega t$ also exist as can be seen by substituting any of the solutions

$$x = A\, e^{i\omega t},$$

$$x = A \sin \omega t,$$

$$x = A \cos \omega t \tag{2.1.5}$$

into the Newton-Hooke equation for the harmonic oscillator. A general rule for equations of this kind is that the sum or difference of two or more solutions is also a solution. This can be verified by substituting an appropriate sum or difference into the Newton-Hooke equation.

In a *conservative system* in which there is no frictional or similar loss in energy, the force is related to the *potential energy* V by

$$V = -\int f dx; \tag{2.1.6}$$

hence, for the Hooke's law force,

$$V = -\int_0^x -kx dx = \frac{1}{2}kx^2. \tag{2.1.7}$$

The total energy E is *kinetic energy* T plus potential energy V,

$$E = \frac{1}{2}m\left(\frac{dx}{dt}\right)^2 + \frac{1}{2}kx^2, \tag{2.1.8}$$

where $dx/dt$ is a velocity if its direction is specified (a vector) or a speed if the direction is ignored (a scalar).

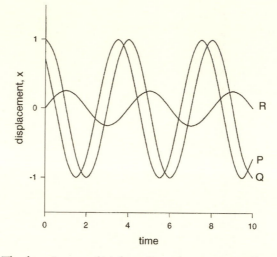

Figure 2.1.2. The function $x = f(t)$ for a linear harmonic oscillator of one mass. Time $t$ is measured in units of $\pi/2\omega$. One period $\tau$ is 4 units on the $t$ axis. The initial condition is $x(0) = 0.707$ distance unit for trajectory P, 1.00 for trajectory Q, and 0.0 distance units for trajectory R. The second initial condition is $\dot{x} = 0.707, 0.0$, and 0.250 speed units for P, Q, and R respectively.

The locus of positions versus time over a specified period $t - t_0$ is a *trajectory* (figure 2.1.2). Any simple harmonic trajectory can be represented by

$$x = A\cos(\omega t + \phi) \qquad (2.1.9\text{a})$$

or by

$$x = A\sin(\omega t + \theta), \qquad (2.1.9\text{b})$$

where $A$ is the maximum amplitude (maximum excursion of the mass) and $\phi$ and $\theta$ are phase angles which locate the sine wave on the time axis. The sine and cosine descriptions of a harmonic oscillator trajectory differ only by a change of $\pi/2$ in the phase angle; hence, as already said, they are equally valid solutions of the Newton-Hooke equation. Integrating a second-order differential equation twice produces two constants of integration. Once we have characterized a frictionless harmonic oscillator by specifying two constants, for example, $A$ and $\phi$, there is nothing more we can say about it. We have used up both of the two constants of integration allowed by the second-order Newton-Hooke differential equation.

## 2.2 Wave Equations: The Vibrating String

Suppose a vibrating guitar string resembles figure 2.2.1a at some instant in time, $t$. Let the displacement from the equilibrium position $u(x, t) = 0$ of some infinitesimal part of the string be $u(x, t)$ on the vertical axis. Each infinitesimal length of the string is undergoing harmonic motion about $u(x, t) = 0$ in response to a vertical force. There is no horizontal motion and no net force in the horizontal direction. The tension $\tau$ at any point of the string is the magnitude of the force acting at that point. The tension is tangential to the string. Because there is no net horizontal force, the $x$ components of the force at any two points of the string must be equal and opposite. The tension on two *adjacent* points, being a scalar, is the same for each.

Figure 2.2.1b shows a tangent line drawn at an arbitrary point on the string $u(x, t)$ between the end points $x = 0$ and $x = L$. Figures 2.2.2a and b show this part of the string in more detail. Let $\alpha$ be the angle between the horizontal and the tangent at $x$. Let $\beta$ be the angle between the horizontal

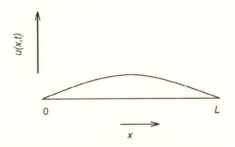

Figure 2.2.1a. The fundamental mode of vibration of a stretched string of length $L$. The horizontal line is the equilibrium position of the string.

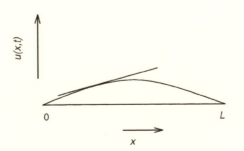

Figure 2.2.1b. The fundamental mode of vibration of a stretched string of length $L$. A line has been drawn tangent to $u(x, t)$ at an arbitrary value of $x$.

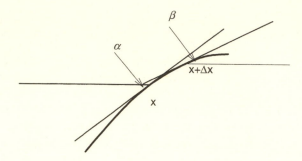

Figure 2.2.2a. Angles $\alpha$, between the horizontal and the tangent at $x$, and $\beta$, between the horizontal and $x + \Delta x$. On this part of the wave, $\alpha > \beta$.

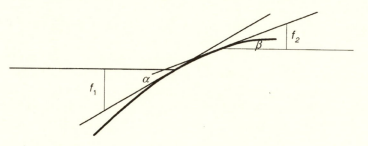

Figure 2.2.2b. Magnitudes of the vertical (transverse) forces $f_1$ and $f_2$ acting on the string at $x$ and $x + \Delta x$.

and another tangent drawn at a slightly different point, $x + \Delta x$. The points are not quite adjacent because $\Delta x$ is finite.

We would like to know how $u(x, t)$ changes with $x$ and $t$, that is, we would like to find a partial differential equation relating $u, x$, and $t$. The magnitude of the vertical force $f_2$ on the string at $x + \Delta x$ differs from the magnitude of the vertical force $f_1$ at $x$ by

$$\tau_2 \sin \beta - \tau_1 \sin \alpha = f_2 - f_1. \tag{2.2.1}$$

The magnitude of the force on the string segment of length $\Delta x$ and unit thickness is, by Newton's second law, $f = ma$, where $m$ is the mass of the segment (density $\rho$ times $\Delta x$) and the transverse acceleration is $a = \partial^2 u(x, t) / \partial t^2$, so

$$\tau_2 \sin \beta - \tau_1 \sin \alpha = \rho \Delta x \frac{\partial^2 u(x, t)}{\partial t^2}. \tag{2.2.2}$$

Because the horizontal components of the force at $x$ and $x + \Delta x$ are equal and opposite, their magnitudes are equal:

$$\tau_2 \cos \beta = \tau_1 \cos \alpha = \tau$$

where $\tau$ is the tension of the unplucked string. We can divide equation 2.2.2 by the horizontal component, which gives

$$\frac{\tau_2 \sin \beta}{\tau_2 \cos \beta} - \frac{\tau_1 \sin \alpha}{\tau_1 \cos \alpha} = \frac{\rho \Delta x}{\tau} \frac{\partial^2 u(x,t)}{\partial t^2} \qquad (2.2.3)$$

or

$$\tan \beta - \tan \alpha = \frac{\rho \Delta x}{\tau} \frac{\partial^2 u(x,t)}{\partial t^2}. \qquad (2.2.4)$$

The tangents are the slopes at $x$ and $x + \Delta x$,

$$\left( \frac{\partial u(x,t)}{\partial x} \right)_{x+\Delta x} - \left( \frac{\partial u(x,t)}{\partial x} \right)_{x} = \frac{\rho \Delta x}{\tau} \frac{\partial^2 u(x,t)}{\partial t^2} \qquad (2.2.5)$$

or

$$\frac{\left( \dfrac{\partial u(x,t)}{\partial x} \right)_{x+\Delta x} - \left( \dfrac{\partial u(x,t)}{\partial x} \right)_{x}}{\Delta x} = \frac{\rho}{\tau} \frac{\partial^2 u(x,t)}{\partial t^2}, \qquad (2.2.6)$$

but the difference in magnitude between two slopes with respect to a change in $x$ approaches the curvature $\partial^2 u(x,t)/\partial x^2$ as $\Delta x$ approaches zero (that is, as the points become more nearly adjacent). This yields

$$\frac{\partial^2 u(x,t)}{\partial x^2} = \frac{\rho}{\tau} \frac{\partial^2 u(x,t)}{\partial t^2} \qquad (2.2.7)$$

which is the wave equation we sought.

## 2.3 Wave Motion

A known tension can be placed on a uniform string, for example, by using it to suspend a weight from a fixed beam. When this is done and a sharp transverse pulse is given near one end of the string, the pulse travels along the string with a pulse velocity of magnitude v. Experimentally, one finds that

$$v = \sqrt{\frac{\tau}{\rho}} \qquad (2.3.1)$$

where $\tau$ is the magnitude of the initial tension on the wire and $\rho$ is its uniform density. (Tension imparted by the pulse is transverse and does not affect the coaxial component of the tension.) If a train of waves is

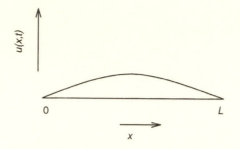

Figure 2.3.1. The fundamental mode of vibration of a stretched string of length L. The function $u(x, t)$ describes one-half of a sine wave.

created by sequential pulses, the train travels at the same speed as a single pulse. For the purpose of studying electromagnetic radiation, we shall be especially interested in sinusoidal wave trains. With the substitution $v^2 = \tau/\rho$ from equation 2.3.1, we can write the wave equation

$$\frac{\partial^2 u(x, t)}{\partial x^2} = \frac{1}{v^2} \frac{\partial^2 u(x, t)}{\partial t^2}. \tag{2.3.2}$$

If the speed and the wavelength of a sinusoidal wave train are both known, we also know the frequency

$$v = \frac{v}{\lambda} \tag{2.3.3}$$

where $v$, the magnitude of the velocity $\mathbf{v}$, is the speed (be careful to distinguish the speed $v$ from the frequency $v$) and $\lambda$ is the wavelength. The frequency in equation 2.3.3 is expressed in hertz. One hertz is a complete cycle per second. Because of the analogy between circular motion and harmonic motion, frequencies are often expressed as $\omega$, measured in radians per second. The conversion is $\omega = 2\pi v$. In electromagnetic theory, confusion between $v$ and $v$ is avoided by calling the speed of electromagnetic radiation $c$. In a vacuum, $c$ is a constant, $c = 2.998 \times 10^8 \text{ m s}^{-1}$. In dense media such as water or glass, the speed of light is considerably less than $c$ (see chapter 9).

Figure 2.3.1 shows the lowest frequency (longest wavelength) that can be produced by striking a string with fixed ends that is $L$ meters long. The wave shown is one-half of a sine wave, so the permissible standing waves on this string have $\lambda = 2L, 2L/2, 2L/3, \ldots$ corresponding to frequencies of $v_0, v_1, v_2, \ldots$. A guitar string at 440 Hz (A natural) would have overtones at 880, 1320, $\ldots$ Hz. An infinite number of frequencies is possible for the vibrating string but, for real strings, overtones diminish rapidly and only a few contribute to the perceived tone of the instrument.

At higher frequencies, inertial mass and resistance to bending damp the oscillation. A real string is not a conservative system.

## 2.4 Solving the Wave Equation: Separation of Variables

Assume that the wave function $u(x,t)$ can be separated into two parts, one a function of $x$ and one a function of $t$:

$$u(x,t) = X(x)T(t). \tag{2.4.1}$$

We verify this assumption by showing that physically plausible results arise from the solutions of the wave equations that are obtained using it. Under the assumption of separability, the wave equation

$$\frac{\partial^2 u(x,t)}{\partial x^2} = \frac{1}{v^2}\frac{\partial^2 u(x,t)}{\partial t^2} \tag{2.4.2}$$

can be written

$$T(t)\frac{d^2 X(x)}{dx^2} = \frac{1}{v^2}X(x)\frac{d^2 T(t)}{dt^2}. \tag{2.4.3}$$

If we divide by $u(x,t) = X(x)T(t)$, we get

$$\frac{1}{X(x)}\frac{d^2 X(x)}{dx^2} = \frac{1}{v^2 T(t)}\frac{d^2 T(t)}{dt^2}. \tag{2.4.4}$$

The variables $x$ and $t$ are independent. If we hold $t$ constant, the left-hand side of the separated equation is equal to a constant no matter what the value of $x$. If the left-hand side of the equation is equal to a constant, the right-hand side is equal to the same constant, no matter what the value of $t$. Let us call the constant $-\beta^2$:

$$\frac{1}{X(x)}\frac{d^2 X(x)}{dx^2} = -\beta^2 \tag{2.4.5}$$

and

$$\frac{1}{v^2 T(t)}\frac{d^2 T(t)}{dt^2} = -\beta^2. \tag{2.4.6}$$

That is,

$$\frac{d^2 X(x)}{dx^2} + \beta^2 X(x) = 0 \tag{2.4.7}$$

and

$$\frac{d^2 T(t)}{dt^2} + \beta^2 v^2 T(t) = 0, \tag{2.4.8}$$

where $v$ is the speed of the wave and $-\beta^2$ is called a *separation constant*. Let us call these two equations the $X(x)$ equation and the $T(t)$ equation. The constant $-\beta^2$ is written as a square to maintain an analogy with $\omega^2$ in the harmonic oscillator equation. The sign of $-\beta^2$ is chosen to bring about an oscillatory solution such as we found for the physical system of a real harmonic oscillator. If $\beta$ were positive, the solution would not be oscillatory.

## 2.5 The Time-Independent Wave Equation $X(x)$

The Newton-Hooke equation for the harmonic oscillator

$$\frac{d^2x}{dt^2} = -\frac{k}{m}x \tag{2.5.1}$$

can be written

$$\frac{1}{x(t)}\frac{d^2x(t)}{dt^2} = -\frac{k}{m} = -\omega^2 \tag{2.5.2}$$

or

$$\frac{d^2x(t)}{dt^2} + \omega^2x(t) = 0 \tag{2.5.3}$$

where $\omega = \sqrt{k/m}$ is the frequency of harmonic oscillation. The sign of $\omega^2$ is in agreement with the separation constant $-\beta^2$ in the $X(x)$ equation, so as to bring about an oscillatory solution.

In treating the harmonic oscillator, we were led to many solutions, among which are

$$x(t) = Ae^{i\omega t},$$

$$x(t) = A\sin\omega t,$$

$$x(t) = A\cos\omega t,$$
$$x(t) = A\cos(\omega t + \phi), \tag{2.5.4a}$$

and

$$x(t) = A\cos\omega t \pm B\sin\omega t, \tag{2.5.4b}$$

where the sum and difference of two solutions is a solution by the *super-position principle*. A mathematical statement of the superposition principle is that, *if $u_1$ and $u_2$ are solutions of a linear homogeneous partial*

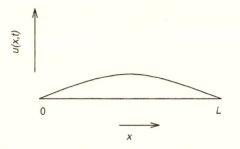

Figure 2.5.1. Vibration of a string fixed at each end (fundamental mode). The maximum of the displacement function (antinode) does not move.

*differential equation, then*

$$u = c_1 u_1 + c_2 u_2 \tag{2.5.5}$$

*is also a solution*, where $c_1$ and $c_2$ are arbitrary constants.

In the same way, the $X(x)$ equation

$$\frac{d^2 X(x)}{dx^2} + \beta^2 X(x) = 0$$

leads to many solutions, including

$$X(x) = A \cos \beta x \pm B \sin \beta x,$$
$$X(x) = B \sin(\beta x + \phi), \tag{2.5.6a}$$

or

$$X(x) = B \cos(\beta x + \theta), \tag{2.5.6b}$$

where the only difference between the sin and cos solutions is the phase angle (see figure 2.1.2). In the solution

$$X(x) = B \sin \beta x \tag{2.5.6c}$$

we have arbitrarily set the phase angle $\phi = 0$. This amounts to the boundary condition $X(x) = 0$ at $x = 0$ in figure 2.5.1. Two boundary conditions are allowed for a second-order differential equation. The second boundary condition that must be satisfied by a string fixed at both ends is $X(x) = B \sin \beta x = 0$ at $x = L$. The amplitude constant $B$ cannot be zero because, if it were, $X(x)$ would be zero for the entire length of the string (a trivial solution involving no vibration at all). Therefore $\sin \beta x = 0$ at $x = L$.

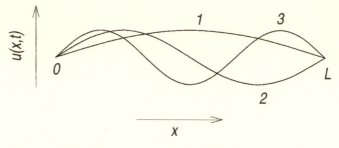

Figure 2.5.2. Standing vibratory waves (fundamental, first, and second overtones) in a string fixed at each end. The integers $n = 1, 2, 3, \ldots$ correspond to different modes of vibration.

The condition $\sin \beta x = 0$ is also satisfied by $\sin n\pi = 0$, where $n$ is a positive integer (see figure 2.1.2). Now, if

$$B \sin \beta L = B \sin n\pi, \quad n = 1, 2, 3, \ldots, \tag{2.5.7}$$

then

$$\beta L = n\pi,$$
$$\beta = \frac{n\pi}{L}, \quad n = 1, 2, 3, \ldots, \tag{2.5.8}$$

and the second boundary condition leads to

$$X(x) = B \sin \frac{n\pi x}{L}, \quad n = 1, 2, 3, \ldots, \tag{2.5.9}$$

for the time-independent part of the wave equation. The wave does not move; hence it is called a *standing wave*. The integers $n = 1, 2, 3, \ldots$ describe the fundamental *mode* and the first, second, $\ldots$ *overtones* of the vibration. These modes are called the first, second, third, $\ldots$ *harmonics*. The mode of vibration with $n = 2$ is the second harmonic but it is the first overtone. The mode with $n = 3$ is the third harmonic but the second overtone, and so on.

The term *node* (not to be confused with mode) is used to designate the point at which the wave function crosses the $u(x, t) = 0$ axis for a standing wave. For example, the fundamental mode of vibration in figure 2.5.2 has only two nodes, one at each end of the string. Higher harmonics have, sequentially, more nodes. The number of nodes for each harmonic is $n + 1$. One sometimes sees the term *internal nodes*, implying exclusion of the two terminal nodes, which must, obviously, be the same for all harmonics. The number of internal nodes is $n - 1$.

Remember that the number of waves per second times the wavelength (distance covered) of each wave is its speed in distance per second,

$v\lambda = v = \omega\lambda/2\pi$. The more nodes a wave function has, the shorter the wavelength and the higher the frequency for a given speed of propagation v. The energy imparted to a string by initially striking or plucking it is quickly distributed among the fundamental and overtones, but the amount of energy retained by the overtones diminishes rapidly with increasing $n$, which is why the tone we hear is dominated by the fundamental. Thus the vertical axis in figure 2.5.2 is an energy axis. Each new harmonic increases the energy of the vibration on the vertical axis.

## 2.6  The Time-Dependent Wave Equation $T(t)$

For the time-dependent equation, we have

$$\frac{d^2 T(t)}{dt^2} + \beta^2 v^2 T(t) = 0. \tag{2.6.1}$$

This wave equation is quite analogous to the Newton-Hooke equation, describing the variation of $T(t)$ with time as the independent variable. The function $T(t)$ oscillates harmonically with $\beta^2 v^2$ as the *fundamental frequency* $\omega_1$. (The notation $\omega_0$ is often used for the fundamental frequency but we shall use $\omega_1$ to preserve correspondence between $\omega_n$ and $n\pi x/L$.) The possibility of higher harmonics (overtones) gives rise to a *spectrum of frequencies* $\omega_n$, of which the fundamental frequency is the lowest.

A general solution of the time-dependent wave equation is

$$T(t) = C \cos \omega_n t + D \sin \omega_n t, \quad \omega_n = \beta v = n\pi v/L. \tag{2.6.2}$$

There are no boundary conditions analogous to those for the standing wave, but we already know from the standing wave that $\beta = n\pi/L$, $X(x) = B \sin(n\pi x/L)$, and from the separation of variables equation we have

$$u(x,t) = X(x)T(t); \tag{2.6.3}$$

hence

$$u(x,t) = B \sin \frac{n\pi x}{L} (C \cos \omega_n t + D \sin \omega_n t) \tag{2.6.4}$$

where $X(x) = $ is one of the solutions that we can select from those given as equations 2.5.4 with the slight modification of $\omega_n$ for $\omega$ to give the entire energy spectrum. Collecting constants, the complete equation is

$$u(x,t) = (E \cos \omega_n t + F \sin \omega_n t) \sin \frac{n\pi x}{L}. \tag{2.6.5}$$

By the superposition principle, the sum of solutions for different values of $n$ is also a solution,

$$u(x, t) = \sum_{n=1}^{\infty} (E \cos \omega_n t + F \sin \omega_n t) \sin \frac{n\pi x}{L}. \qquad (2.6.6)$$

An equivalent form is

$$u(x, t) = \sum_{n=1}^{\infty} G_n \cos(\omega_n t + \phi_n) \sin \frac{n\pi x}{L} = \sum_{n=1}^{\infty} u_n(x, t). \qquad (2.6.7)$$

Each $u(x, t)$ in $\sum_{n=1}^{\infty} u_n(x, t)$ is a *normal mode* and the sum $\sum_{n=1}^{\infty} u_n(x, t)$ is the *superposition* of normal modes.

## 2.7 Superpositions

The fundamental of a standing wave is

$$u(x, t) = G_1 \cos(\omega_1 t) \sin \frac{\pi x}{L} \qquad (2.7.1)$$

where $\phi = 0$ and $n = 1$. We have, from 2.6.2, $\omega_1 = \pi v / L$; hence the frequency in hertz is

$$\frac{\omega_1}{2\pi} = \frac{v}{2L} \qquad (2.7.2)$$

and $\omega_1 t = \pi v t / L$, which leads to

$$u(x, t) = G_1 \sin \frac{\pi x}{L} \cos(\omega_1 t) = G_1 \sin \frac{\pi x}{L} \cos \frac{\pi v t}{L}. \qquad (2.7.3)$$

From the trigonometric identity,

$$\sin \xi \cos \zeta = \frac{1}{2} \sin(\xi + \zeta) + \frac{1}{2} \sin(\xi - \zeta), \qquad (2.7.4)$$

we have

$$u(x, t) = \frac{G_1}{2} \sin \left( \frac{\pi x}{L} + \frac{\pi v t}{L} \right) + \frac{G_1}{2} \sin \left( \frac{\pi x}{L} - \frac{\pi v t}{L} \right) \qquad (2.7.5)$$

$$= \frac{G_1}{2} \sin \left[ \frac{\pi}{L} (x + vt) \right] + \frac{G_1}{2} \sin \left[ \frac{\pi}{L} (x - vt) \right]. \qquad (2.7.6)$$

For the fundamental mode, the wavelength of a standing wave is $\lambda = 2L$; hence

$$u(x, t) = \frac{G_1}{2} \sin \left[ \frac{2\pi}{\lambda} (x + vt) \right] + \frac{G_1}{2} \sin \left[ \frac{2\pi}{\lambda} (x - vt) \right]. \qquad (2.7.7)$$

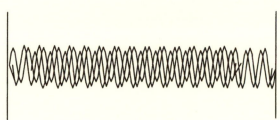

Figure 2.7.1. Destructive interference occurs when boundary conditions are not satisfied.

This is a superposition of two *running* waves $\longrightarrow$ and $\longleftarrow$. (If a wave is not standing, we say it is running.) We can think of these terms as describing a running wave $\longrightarrow$ that strikes a barrier at $L$ and is reflected back $\longleftarrow$ toward $x = 0$. If the boundary conditions $u = 0$ at $x = 0$ and $L$ are satisfied, the wave is undamped and, if it were not for frictional losses, it would last forever. If the boundary conditions are not satisfied, *destructive interference takes place and the wave disappears in finite time.*

Figure 2.7.1 shows destructive interference as it occurs between waves that are bouncing back and forth between the right and left boundaries. The positive amplitudes cancel the negative amplitudes at each point on the wave, yielding a zero sum.

## 2.8  A Two-Dimensional Wave Equation

The two-dimensional wave equation satisfied by a vibrating thin sheet of dimensions $x_1 \times x_2$ (see figure 2.8.1),

$$\frac{\partial^2 u(x_1, x_2, t)}{\partial x_1^2} + \frac{\partial^2 u(x_1, x_2, t)}{\partial x_2^2} = \frac{1}{v^2} \frac{\partial^2 u(x_1, x_2, t)}{\partial t^2}, \qquad (2.8.1)$$

has $u(x_1, x_2, t)$ as a function of space $x_1, x_2$ and time $t$.

As in the one-dimensional wave equation, the two-dimensional surface equation can be separated into a space part and a time part. The separability assumption

$$u(x_1, x_2, t) = F(x_1, x_2)T(t) \qquad (2.8.2)$$

leads to

$$\frac{1}{F(x_1, x_2)} \left[ \frac{\partial^2 F(x_1, x_2)}{\partial x_1^2} + \frac{\partial^2 F(x_1, x_2)}{\partial x_2^2} \right] = \frac{1}{v^2 T(t)} \frac{d^2 T(t)}{dt^2}. \qquad (2.8.3)$$

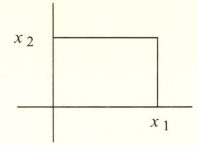

Figure 2.8.1. A thin sheet capable of vibrating in the directions perpendicular to $x_1$ and $x_2$.

The parts are identically equal only if

$$\frac{\partial^2 F(x_1, x_2)}{\partial x_1^2} + \frac{\partial^2 F(x_1, x_2)}{\partial x_2^2} + \beta^2 F(x_1, x_2) = 0 \qquad (2.8.4)$$

and

$$\frac{d^2 T(t)}{dt^2} + \beta^2 v^2 T(t) = 0, \qquad (2.8.5)$$

where $-\beta^2$ is the separation constant.

## 2.9 The Time-Independent Wave Functions $X(x_1)$ and $X(x_2)$

The space part can be separated a second time to obtain an $x_1$ part and an $x_2$ part. We assume that $F(x_1, x_2) = X(x_1)X(x_2)$, which leads to

$$\frac{1}{X(x_1)} \frac{d^2 X(x_1)}{dx_1^2} = -\rho^2 \qquad (2.9.1)$$

and

$$\frac{1}{X(x_2)} \frac{d^2 X(x_2)}{dx_2^2} = -\sigma^2, \qquad (2.9.2)$$

where the relation among separation constants is $\rho^2 + \sigma^2 = \beta^2$. These equations yield solutions

$$X(x_1) = A \cos \rho x_1 + B \sin \rho x_1 \qquad (2.9.3)$$

and

$$X(x_2) = C \cos \sigma x_2 + D \sin \sigma x_2. \qquad (2.9.4)$$

The boundary conditions at $x_1 = x_2 = 0$ force $A = 0$ and $C = 0$. At $x_1 = L_1$ and $x_2 = L_2$,

$$\rho L_1 = n_1 \pi \quad \text{and} \quad \sigma L_2 = n_2 \pi, \quad n_1 = 1, 2, 3, \ldots, \quad n_2 = 1, 2, 3, \ldots,$$

whence

$$\beta = \pi \left( \frac{n_1^2}{L_1^2} + \frac{n_2^2}{L_2^2} \right)^{1/2}. \tag{2.9.5}$$

Now there are two integers $n_1$ and $n_2$ which determine the nodal properties of the standing vibration in two dimensions.

The time-dependent equation yields

$$T(t) = E \cos \omega t + F \sin \omega t$$

or, using the phase angle $\phi$,

$$T(t) = G \cos(\omega t + \phi), \tag{2.9.6}$$

where

$$\omega = \mathsf{v}\beta = \mathsf{v}\pi \left( \frac{n_1^2}{L_1^2} + \frac{n_2^2}{L_2^2} \right)^{1/2}. \tag{2.9.7}$$

The complete solution is a double sum,

$$u(x_1, x_2, t) = \sum_{n_1=1}^{\infty} \sum_{n_2=1}^{\infty} H \cos(\omega t + \phi) \sin \frac{n_1 \pi x}{L_1} \sin \frac{n_2 \pi x}{L_2}$$

$$= \sum_{n_1=1}^{\infty} \sum_{n_2=1}^{\infty} u(x_1, x_2, t). \tag{2.9.8}$$

The motion has fundamental modes of vibration, harmonics, nodes, and antinodes as in one-dimensional wave motion. Waves perpendicular to one another on a square surface lead to the possibility of *degeneracy*. If $L_1 = L_2 = L$, different modes of motion in the $x_1$ and $x_2$ directions can lead to the same frequency $\omega$:

$$\omega = \frac{\mathsf{v}\pi}{L} \left( n_1^2 + n_2^2 \right)^{1/2}. \tag{2.9.9}$$

For example, the combinations $n_1 = 1$, $n_2 = 2$ and $n_1 = 2$, $n_2 = 1$, although they correspond to distinctly different vibrations on the surface pictured in figure 2.9.1, lead to the same $\omega$ and hence to the same total energy of vibration.

The relation between $\omega$ and the integers $n_1$ and $n_2$ shows that the higher the integers, the higher $\omega$, and hence the higher the energy of the vibration.

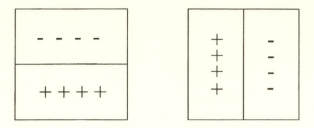

Figure 2.9.1. Degenerate modes of vibration on a square planar surface. The − surface goes down into the plane of the page and the + surface comes out of the plane of the page. The straight lines are nodal lines.

We shall want, ultimately, to know the energy density for vibrations within a selected geometry as a function of the frequency of the waves that satisfy the boundary conditions imposed by the geometry.

## 2.10 A Three-Dimensional Wave Equation

Extension to three dimensions yields an equation that is analogous to the two-dimensional case. Suppose vibrations occur in a cubic solid. (Other geometries work, but a cube gives simpler equations.) If destructive interference is not to occur, all three components of the waves must satisfy the boundary conditions imposed by the cube. The perpendicular components $x_1, x_2,$ and $x_3$ must have nodes at each surface of the cube (figure 2.10.1).

The wave equation in three dimensions,

$$\frac{\partial^2 u(x_1, x_2, x_3, t)}{\partial x_1^2} + \frac{\partial^2 u(x_1, x_2, x_3, t)}{\partial x_2^2} + \frac{\partial^2 u(x_1, x_2, x_3, t)}{\partial x_3^2}$$

$$= \frac{1}{v^2} \frac{\partial^2 u(x_1, x_2, x_3, t)}{\partial t^2}, \tag{2.10.1}$$

is separated in the usual way, leading to three separation constants

$$\rho^2 + \sigma^2 + \tau^2 = \beta^2.$$

As in the two-dimensional case, $\omega = v\beta$. Imposition of boundary conditions on the components of the waves in three dimensions leads to

$$\omega = v\beta = v\pi \left( \frac{n_1^2}{L_1^2} + \frac{n_2^2}{L_2^2} + \frac{n_3^2}{L_3^2} \right)^{1/2} \tag{2.10.2a}$$

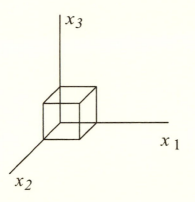

Figure 2.10.1. Each face of a cubic box establishes a nodal plane for a standing wave.

or

$$\omega = \frac{v\pi}{L} \left(n_1^2 + n_2^2 + n_3^2\right)^{1/2}, \tag{2.10.2b}$$

where $L^2$ can be factored out because the edges of the cube are equal in length. By the Pythagorean theorem, $(n_1^2 + n_2^2 + n_3^2)^{1/2} = n$ in a number space of $n_1, n_2$, and $n_3$ measured along orthogonal Cartesian coordinates; hence

$$\omega = \frac{v\pi}{L} n. \tag{2.10.2c}$$

We shall call this space the *quantum number space* as distinct from the $x_1, x_2, x_3$ space. If $n_1, n_2$, and $n_3$ are very large, $n$ approaches the radius of a sphere in the three-dimensional space spanned by $n_1, n_2$, and $n_3$. This fact will be useful to us very soon.

## 2.11 The Boltzmann Distribution

Originally derived on the assumption that energy is continuous, the Boltzmann distribution gives a number density $dn/d\varepsilon$ of particles in an infinitesimal energy interval $\varepsilon + d\varepsilon$,

$$\frac{dn}{d\varepsilon} = A e^{-\varepsilon/kT}. \tag{2.11.1}$$

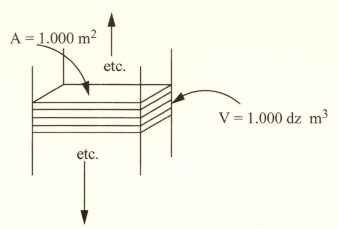

Figure 2.11.1. A column of air segmented into thin $1.000\,\text{m}^2$ slices. The vertical dimension is $z$.

If, as is reasonable, we take all our infinitesimals $d\varepsilon$ to be the same size, the proportionality constant can be made to cancel:

$$\frac{\dfrac{dn_1}{d\varepsilon}}{\dfrac{dn_0}{d\varepsilon}} = \frac{dn_1}{dn_0} = \frac{Ae^{-\varepsilon_1/kT}}{Ae^{-\varepsilon_0/kT}} = e^{-(\varepsilon_1-\varepsilon_0)/kT}.$$

Since the zero of energy can be specified arbitrarily, let $\varepsilon_0 = 0$ and $\varepsilon_1 = \varepsilon$. Now the ratio of number densities is

$$\frac{dn_1}{dn_0} = e^{-\varepsilon/kT}. \tag{2.11.2}$$

An example is the *barometric equation*, in which we consider two infinitesimal slices of a vertical column of air shown in figure 2.11.1. Pressure is force per unit area. The pressure exerted by many particles is proportional to their number because each particle makes its distinct contribution to the force. The pressure in each slice is proportional to the number density in the slice, provided that the temperature is constant along the vertical column, so that

$$\frac{p}{p_0} = e^{-mgh/kT} \tag{2.11.3}$$

where $mgh$ is the potential energy of particles of mass $m$ at a height $h$ above $h = 0$ (sea level) and $g = 9.807\,\text{m}\,\text{s}^{-2}$ is the acceleration due to gravity at sea level. Because pressure is a collective effect of particle impact, the actual number of particles $n$ in a narrow interval of altitude at $h$ is

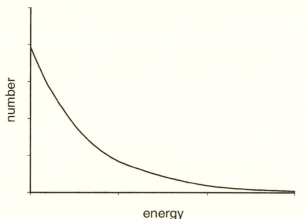

Figure 2.11.2. The Boltzmann function in $n - \varepsilon$ space.

proportional to the number $n_0$ in a similar interval at the defined zero $h = 0$,

$$\frac{n}{n_0} = e^{-mgh/kT} \qquad (2.11.4a)$$

or

$$n = n_0 e^{-mgh/kT} \qquad (2.11.4b)$$

where $k$ is Boltzmann's constant (section 2.14).

The Boltzmann distribution for continuous $\varepsilon$ defines a curve in *number-energy space* (figure 2.11.2). If we recognize that $n$ is a measure of pressure and $\varepsilon$ is a measure of altitude, such a curve describes the change in atmospheric pressure with altitude (complicated by variations in $T$). Just as ordinary density is weight per unit volume, say $kg\,m^{-3}$, we shall often have cause to speak of *number densities* as the number of particles per unit volume, $n\,m^{-3}$, or, in a one-dimensional number-energy space, as the number of particles per unit of energy.

## 2.12 Degrees of Freedom

The idea of degrees of freedom is conveyed nicely by a consideration of the energy of a classical ideal gas. By definition, the potential energy of interparticle interaction in an ideal gas is negligible and the gravitational potential energy within a container is constant provided that the vertical dimension is small with respect to the radius of the earth. The only energy of ideal gas particles is kinetic. Kinetic energy is energy of motion. Collisions in an ideal gas are elastic, so particles are free to exchange energy. In three-dimensional space (3-space), the kinetic energy of a particle can be broken up into a contribution from each dimension of the motion. For the average energy of a large number of particles, there is nothing to make us prefer one dimension over the other two. There are three *degrees of freedom* and energy is *partitioned equally* among them.

## 2.13  Kinetic Energy per Degree of Freedom

Let a mole of noninteracting particles be confined to a cubic container $L$ on an edge. For the moment, concentrate on one particle. When a single particle with a component in the $x$ direction collides with one of the two container walls perpendicular to the $x$ direction, the $x$ component of its momentum, $p_x = mv_x$, changes from $p_x$ to $-p_x$. The total momentum change is $2p_x$, having a scalar magnitude of $\tilde{p}_x = 2mv_x$. (Note that we are using italic $v$ for the particle speed now that there is no danger of confusing it with $v$.)

In order to collide a second time with the same wall, the particle must travel to the opposite wall and back again, a distance of $2L$. The frequency of collisions with the wall perpendicular to the $x$ component of its motion is the $x$ component of its speed divided by the distance it must travel:

$$\text{freq} = \frac{v_x}{2L}. \tag{2.13.1}$$

The magnitude of the force $f$ transmitted to the wall is the momentum change per collision times the number of collisions:

$$f = 2\tilde{p}_x \frac{v_x}{2L} = 2mv_x \frac{v_x}{2L} = m\frac{v_x^2}{L}. \tag{2.13.2}$$

Pressure $p$ is force per unit area and the area of one wall of the container is $L^2$; hence

$$p = \frac{f}{L^2} = \frac{mv_x^2}{L}\left(\frac{1}{L^2}\right) = \frac{mv_x^2}{L^3} = \frac{mv_x^2}{V} \tag{2.13.3}$$

per particle, or

$$pV = mv_x^2 \tag{2.13.4}$$

per particle, where $V = L^3$ is the volume of the container.

By Pythagoras's theorem, the length of the velocity vector is related to its components as

$$v^2 = v_x^2 + v_y^2 + v_z^2.$$

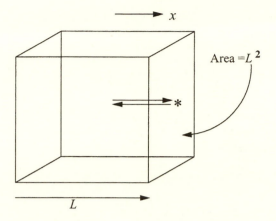

Figure 2.13.1. A gas molecule in a cubic container. Upon executing a perfectly elastic collision * with a wall, the change in momentum is $mv - (-mv) = 2mv$.

The *root mean square* average of $N$ squared speeds for a large collection of particles moving randomly in the container is defined as $v_{rms}^2$:

$$\frac{\sum\limits_{i=1}^{N} v_i^2}{N} \equiv v_{rms}^2 \tag{2.13.5}$$

or

$$v_{rms} = \sqrt{v_{rms}^2} = \sqrt{\frac{\sum\limits_{i=1}^{N} v_i^2}{N}}. \tag{2.13.6}$$

Because $N$ is large, on average,

$$v_x^2 = v_y^2 = v_z^2.$$

Assuming free exchange of energy through particle collisions, no direction is preferred over any other, that is,

$$v_{rms}^2 = 3v_x^2$$

or

$$v_x^2 = \frac{1}{3}v_{rms}^2. \tag{2.13.7}$$

The pressure-volume product due to $N$ molecules impacting on the right-hand wall of the cube perpendicular to the $x$ coordinate is

$$pV = Nmv_x^2 = \frac{1}{3}Nmv_{rms}^2. \tag{2.13.8}$$

This equation gives the pressure on the other faces of the cube too, because pressure is constant within the container. Multiplying and dividing by 2 leads to

$$pV = \frac{2}{3}N\left(\frac{1}{2}mv_{rms}^2\right) = \frac{2}{3}N\ \bar{\varepsilon}_{kin} \qquad (2.13.9)$$

where the root-mean-square kinetic energy is $\frac{1}{2}mv_{rms}^2$ and $\bar{\varepsilon}_{kin}$ is the kinetic energy for a representative or average particle in the ensemble. For a large number of particles, $N$,

$$pV = \frac{2}{3}N\ \bar{\varepsilon}_{kin} = \frac{2}{3}\bar{E}_{kin} \qquad (2.13.10)$$

where $\bar{E}_{kin}$ is the average value of the total energy of all the particles. It turns out, from statistical arguments, that fluctuations in energy away from $\bar{E}_{kin}$ are vanishingly small for a very large number of particles. Therefore, we drop the notation $\bar{E}_{kin}$ and simply regard $E_{kin}$ as the total kinetic energy. We can select the Avogadro number as the number of particles in our ensemble, $N = N_A$, and, since we know that $pV = RT$ for one mole of an ideal gas, we have

$$E_{kin} = \tfrac{3}{2}PV = \tfrac{3}{2}RT \qquad (2.13.11)$$

where, for $N = N_A$, $E_{kin}$ is the *molar* energy in *joules*, J. The important principle emerges that the *average kinetic energy per particle* $\bar{\varepsilon}_{kin}$ is $3RT/2N_A$ or $RT/2N_A$ *per particle per degree of freedom.*

## 2.14 Boltzmann's Constant

From

$$\bar{\varepsilon}_{kin} = \frac{3}{2}\frac{R}{N_A}T \qquad (2.14.1)$$

we can find the important constant

$$\frac{R}{N_A} = \frac{8.3145}{6.022 \times 10^{23}} = 1.381 \times 10^{-23}\,\mathrm{J\,K^{-1}} = k_B, \qquad (2.14.2)$$

that is,

$$\bar{\varepsilon}_{kin} = \tfrac{3}{2}k_B T \qquad (2.14.3)$$

where $k_B$ is the universal gas constant *per particle*. From this point on, we shall denote this constant $k_B$. Written in terms of $k_B$ and equally partitioned among three degrees of freedom the *kinetic energy per degree of freedom is* $\frac{1}{2}k_BT$. It is to be emphasized that everything said here derives from classical mechanics, not quantum mechanics.

## 2.15 The Translational Energy

The energy calculated in this way is frequently called the *translational energy* because it is calculated from a model consisting of point masses executing only translational motion in 3-space as distinct from extended molecular structures, which may also execute rotational or vibrational motion.

To confirm this, we can look to the experimental thermodynamics of the noble (monatomic) gases which are, to a good approximation, non-interacting particles. For a small temperature rise,

$$\frac{dE}{dT} \equiv C_V$$

where $E$ is the molar energy and $C_V$ is the *molar heat capacity at constant volume*. If it is true that $E_{\text{ideal gas}} = \frac{3}{2}RT$ for one mole, then $E = \frac{3}{2}(8.3145)T = 12.472T$ in joules, also for one mole of an ideal gas. For experimental verification, we ask whether the constant-volume molar heat capacity is $d(12.472T)/dT = 12.472$ J K$^{-1}$mol$^{-1}$. Helium and neon are examples of real gases that are very nearly ideal and that have no vibratory or rotatory modes of motion. The experimental molar heat capacities at constant volume are

$$C_V(\text{He}) = 12.471 \text{ J K}^{-1} \text{ mol}^{-1},$$
$$C_V(\text{Ne}) = 12.471 \text{ J K}^{-1} \text{ mol}^{-1},$$

giving us a very impressive confirmation of the theoretical value.

## 2.16 The Energy of a Vibrational State Is $k_BT$

We have seen that the average energy of an ideal gas confined to a cubic container is $\frac{1}{2}k_BT$ per particle per degree of freedom. If a system of fixed harmonic oscillators is in the same cubic container along with an ideal gas, its average kinetic energy must be $k_BT$ per degree of freedom per oscillator. If it were not, the system would not be in thermal equilibrium. The difference is that an ideal gas has three degrees of freedom, the $x$, $y$, and $z$ dimensions, while a harmonic oscillator has only one, along the coordinate of vibration. With only one degree of freedom, why is the energy of the harmonic oscillator $k_BT$ rather than $\frac{1}{2}k_BT$? There is only one kind of energy in an ideal gas, kinetic energy of translation, while a harmonic oscillator has two, kinetic energy *and* potential energy of vibration. Classically, the kinetic energy and potential energy of a harmonic oscillator are equal. Each vibrational state contributes $2(\frac{1}{2}k_B T) = k_B T$ to the energy of the system.

Boltzmann argued (in 1871) that a solid element can be regarded as $N$ harmonic oscillators, each with three degrees of freedom, $x$, $y$, and $z$ in Cartesian space. Since $N = N_A$, the Avogadro number for a molar quantity, the molar energy $U$ of a pure solid element should be

$$U = 3N_A k_B T = 3RT \qquad (2.16.1)$$

where $R$ is the molar gas constant, $R = N_A k_B$; hence the molar heat capacity, which is the specific heat per gram times the gram atomic weight, is

$$C = \frac{\partial U}{\partial T} = 3R = 3(8.3145) = 24.9 \, \text{J} \, \text{mol}^{-1},$$

in agreement with the law of Dulong and Petit. (Where an unsubscripted $C$ is used to denote the specific heat, constant volume should be assumed.)

## 2.17 Trouble Brewing

The preceding arguments assume that the oscillator is fixed to a lattice site in a crystal. If a diatomic molecule, which is free to move within a container, also executes harmonic motion along its bond axis, its energy is translational *and* vibrational, $3(\frac{1}{2}k_B T) + 2(\frac{1}{2}k_B T) = \frac{5}{2}k_B T$ per molecule. The expected molar heat capacity, $\frac{5}{2}R = \frac{5}{2}(8.1345) = 20.79 \, \text{J} \, \text{K}^{-1} \, \text{mol}^{-1}$, is found near room temperature for some diatomic molecules in the gaseous state, but not for others:

$$C_V(N_2) = 20.79 \, \text{J} \, \text{K}^{-1} \, \text{mol}^{-1},$$

$$C_V(O_2) = 21.04 \, \text{J} \, \text{K}^{-1} \, \text{mol}^{-1},$$

$$C_V(F_2) = 22.99 \, \text{J} \, \text{K}^{-1} \, \text{mol}^{-1}.$$

One can rationalize a heat capacity that is higher than $20.81 \, \text{J} \, \text{K}^{-1} \, \text{mol}^{-1}$ by invoking the idea that a diatomic molecule rotates about its center of gravity, but addition of a "little bit" to the molar heat capacity violates the rule of $\frac{1}{2}k_B T$ per degree of freedom. *This is a serious problem for classical physics.* We shall see that the problem of anomalous low-temperature heat capacities of crystals is similar to the problem of anomalous heat capacities of diatomic molecules, and when we have the solution to one problem, we shall have the solution to the other.

## PROBLEMS

**2.1.** Show that $x = Ae^{i\omega t}$ is a solution of (satisfies) the Newton-Hooke equation, equation 2.1.3.

**2.2.** Show that the sum of solutions

$$x = A \sin \omega t + B \cos \omega t$$

is a solution to the Newton-Hooke equation.

**2.3.** Referring to figure 2.1.2, how do we know that at $t = 0$, $\dot{x} = 0$ for curve Q?

**2.4.** Taking $m = 1.00$ kg, the displacement in figure 2.1.2 in meters, and $k = 1.00$ N m$^{-1}$ ($\omega = 1.00$ s$^{-1}$), what are the initial potential energies for the mass if it is to follow trajectories P, Q, and R?

**2.5.** Given the conditions in problem 2.4, find the initial speed of the 1.00 kg mass starting out on trajectories P, Q, and R in figure 2.1.2, that is, verify the last sentence in the caption of figure 2.1.2. The maximum excursion of trajectory R is 0.250 m.

**2.6.** Sketch the normal modes of vibration of a square membrane analogous to figure 2.9.1 for $\{n_1 = 1, n_2 = 1\}$, $\{n_1 = 1, n_2 = 2\}$, $\{n_1 = 2, n_2 = 1\}$, $\{n_1 = 2, n_2 = 2\}$, $\{n_1 = 1, n_2 = 3\}$, $\{n_1 = 3, n_2 = 1\}$, $\{n_1 = 3, n_2 = 3\}$. Which are degenerate?

**2.7.** A 2.00 kg mass is suspended from a fixed point by a 6.00 m length of 2.00 mm diameter copper wire. A slight "ping" is struck with a mallet near the fixed end, causing a transverse wave to travel down the wire. How long does it take the wave to get to the 2.00 kg mass? The density of copper is 8920 kg m$^{-3}$.

# Three

## Experimental Background

THEORETICIANS LIKE WIEN AND PLANCK exchanged information with experimentalists like Paschen, Lummer, and Pringsheim in a cooperation that has not always characterized science, particularly German science at the end of the nineteenth century. Late in that century, a remarkable group of men gathered in Berlin to work on the problem of the radiation *spectrum* emitted by a heated *blackbody*, a kind of ideal energy radiator related to real radiation in much the same way that an ideal gas is related to a real gas. This chapter focuses on Kirchhoff's definition of an ideal blackbody, and the experimental successes and failures experienced by those who attempted to determine the spectrum of its emitted electromagnetic energy.

### 3.1 Thermal Radiation in a Chamber or Cavity

Consider two objects A and B, not in contact with one another, existing in an evacuated chamber insulated from the outside world (figure 3.1.1). If at some time one is hotter than the other, at some later time they will have arrived at the same temperature. The hot object will radiate heat to the cool one until they are in thermal equilibrium. As the hot object cools, it emits *thermal radiation*, which is absorbed by the cool object as it warms up. Thermal radiation is electromagnetic radiation, which can be regarded as an oscillation in the electromagnetic field. The "evacuated chamber" cited above does not contain nothing, it contains radiation.

We immediately arrive at a first principle of radiation in an evacuated chamber, one that will be very important as we proceed. We have given no specific properties of objects A and B within the box, stipulating only that they approach thermal equilibrium. It is clear that we can move A and B around in the box, that they may be of different size and geometry, and that we can rotate them, without altering the equilibrium. Also, since we haven't said what the objects are made of, they may be of different materials, with all physical properties unrelated save one: temperature.

The radiation energy within the box, which is itself in equilibrium with A and B, is homogeneously distributed and has a density, radiation energy

Figure 3.1.1. An insulated, evacuated box containing objects A and B.

per unit volume, which we shall call $u = U/V$, where $U$ is the total radiation energy and $V$ is the volume of the box. From the previous argument, $u$ is a function only of the temperature:

$$u = f(T). \qquad (3.1.1)$$

When thermal equilibrium has been reached, it is not reasonable to suppose that emission and absorption of radiation cease. Emission and absorption continue, only their rates are equal (a familiar concept to the chemist by analogy with chemical equilibrium),

$$A \leftrightharpoons B.$$

As long as the temperature is above 0 K and equilibrium exists, A and B will continue to emit and absorb radiation at the same rate.

Now remove objects A and B and surround the walls of the box with a heating mantle in thermal contact with the box so that its temperature can be controlled. Once again, radiation exists within the box after thermal equilibrium has been established at some constant temperature $T > 0$ K. The walls of the box must be emitting and absorbing radiation at the same rate if they are to be at thermal equilibrium (not warming or cooling). The walls of the box now play the role of an "object" within the box. There must be a constant amount of radiation within the box, which is determined only by the temperature. This is quite remarkable. Unlike matter, radiation is not conserved. Radiation can be created or destroyed by warming or cooling the box (figure 3.1.2).

Since radiation is a means of transferring energy, $U \propto T$, we need a proportionality constant between the total energy $U$ and $T$. The kinetic theory of gases shows us that the translational energy is related to temperature as $\frac{1}{2}k_B T$ per degree of freedom, where $k_B$ is Boltzmann's constant. We can regard the submicroscopic structure (atoms) of the walls of our box as consisting of real or virtual harmonic oscillators. The harmonic oscillator in one dimension has one degree of freedom, which contributes to both potential energy and kinetic energy, leading to $2\left(\frac{1}{2}k_B T\right) = k_B T$ of energy per vibrational mode. Thus $k_B$ is the proportionality constant

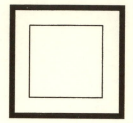

Figure 3.1.2. A thermostated evacuated box. From the point of view of an observer in the box, there is no difference between a perfectly insulated box and a perfectly thermostated box.

sought and

$$U = k_B T \tag{3.1.2}$$

for the one-dimensional classical harmonic oscillator excited by thermal energy.

By the second half of the nineteenth century, experimentalists had demonstrated that both heat and light are parts of the electromagnetic spectrum, differing only in wavelength $\lambda$. Measurements on the heat and light of the sun informed us that energy is not normally emitted or received at one or a few wavelengths but over a broad spectrum from very short to very long wavelengths. Now suppose that, within our evacuated chamber in figure 3.1.1, emission and absorption were exactly the same for objects A and B at thermal equilibrium except for one special wavelength $\lambda_s$. If at this special wavelength object A absorbed heat radiation better than B, then A would become spontaneously hotter than B, violating the condition of equilibrium as well as the second law of thermodynamics. Therefore, the special wavelength $\lambda_s$ does not exist and what is true of the total energy density of the entire electromagnetic spectrum, $u = f(T)$, is true of any *monochromatic* part of it,

$$u_\lambda = f(T). \tag{3.1.3}$$

The spectrum is not flat. If we sample at different $\lambda$ but at constant temperature, we get different densities $u_\lambda$. If we vary both $\lambda$ and $T$, we find $u_\lambda = f(\lambda, T)$. We sometimes write $u_{\lambda,T}$ but constant temperature, once stipulated, is usually left out of the notation. We shall be looking at two different problems: We wish to know how the total radiation density $u$ varies with temperature and how monochromatic radiation density $u_\lambda$ varies with wavelength at some fixed temperature.

## 3.2 Kirchhoff's Law: Absorptivity

Kirchhoff proposed in 1860, as has been richly verified since, that, within a chamber of any geometry, at equilibrium, any two different areas $ds$ of the interior must emit and absorb radiation at equal rates to maintain equilibrium (figure 3.2.1).

Let the *emissivity* of radiant energy emitted from an element $ds$ into the chamber be called $e$, and the *fraction* of radiant energy absorbed from incident radiation of intensity $I(\lambda, T)$ be $a$. The fraction $a$ is near 1.0 if radiation is very well absorbed and near 0 if it is poorly absorbed (reflected or scattered). The energy absorbed is $a\,I(\lambda, T)$, which must be equal to the energy emitted $e$ for the condition of thermal equilibrium to be satisfied:

$$a\,I(\lambda, T) = e,$$
$$I(\lambda, T) = \frac{e}{a}. \tag{3.2.1}$$

For cavity radiation, the intensity (which is related to the energy density through a constant, see section 3.3) is a function only of $\lambda$ and $T$, because $e$ and $a$ are functions of $\lambda$ and $T$ (see above):

$$I(\lambda, T) = \frac{e(\lambda, T)}{a(\lambda, T)}, \tag{3.2.2}$$

which is Kirchhoff's law.

If there is a very small hole of area $ds$ in the chamber (figure 3.2.2) and a very small amount of radiation escapes from it into the rest of the universe, that radiation has been perfectly absorbed by the universe and it will never find its way back into the chamber.
Kirchhoff's law is satisfied and $a = 1$, whence

$$I(\lambda, T) = e(\lambda, T) \tag{3.2.3}$$

From this point on, we shall take Kirchhoff's term *blackbody* to be synonymous with the chamber just described. It is interesting to note that

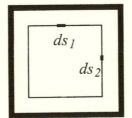

Figure 3.2.1. Two areas on the wall of a thermostated chamber at thermal equilibrium. Different areas $ds_1$ and $ds_2$ are analogous to A and B in figure 3.1.1.

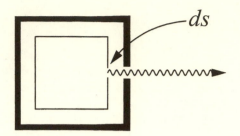

Figure 3.2.2. A thermostated chamber with an infinitesimal radiation leak.

in any real experimental setup the chamber need not be black inside; it is the hole that is black ($a = 1$). The intensity of radiation from a chamber with an infinitesimal aperture is the same as the intensity of radiation within the chamber. Energy density is proportional to intensity of radiation; hence the *shape* of the spectrum of intensity measured at the aperture of a blackbody is the same as the shape of the spectrum of energy density within the box.

## 3.3 The Intensity of Radiation

At this point, we need a clear definition of radiation intensity and a relation between intensity and energy density within the blackbody radiator. First we note that, because of the reciprocal relationship between wavelength and frequency of electromagnetic radiation, notations like $I(\lambda, T)$ and $I(v, T)$ are equivalent. Toward the end of the nineteenth century, notation using frequency became more widely used. Experimentalists tended to express their results in terms of wavelength and theoreticians tended to use frequency. We shall generally follow that tradition, discussing experimental work in terms of $\lambda$ and theoretical work in terms of $v$.

The energy *flux* is the energy passing through unit area in unit time ($J\,m^{-2}\,s^{-1} = watt\,m^{-2} = W\,m^{-2}$). The energy flux due to electromagnetic radiation is given by

$$flux = I(v, T) \cos \theta \, d\Omega \, ds \qquad (3.3.1)$$

(see figure 3.3.1). For a uniformly cylindrical pencil of radiation normal to a very small target area $ds$ inside a cavity, $\cos \theta$ is 1 (figure 3.3.2).

Flux is energy transfer per unit time per unit area:

$$flux = \frac{U}{t\,A} = I(v, T). \qquad (3.3.2)$$

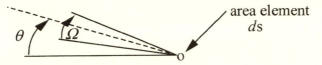

Figure 3.3.1. Intensity of radiation falling from a solid angle $d\Omega$ onto an area element $ds$. The angle between the normal and the incident radiation is $\theta$ so $\cos\theta$ is the horizontal component of the incident radiation. (Kondepudi and Prigogine 1998).

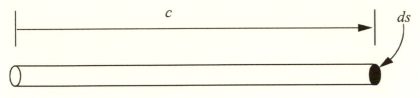

Figure 3.3.2. A uniform pencil of radiation.

For radiation to strike the target in a time interval $t = 1\,\mathrm{s}$, it must be within a distance that electromagnetic radiation can cover in one second. Electromagnetic radiation travels at $c = 2.998 \times 10^8\,\mathrm{m\,s^{-1}}$, so the radiation must be within $c$ meters to strike the target in the allotted time. If we divide both sides of the flux equation by $c$, and note that $cA$ is a length times an area, that is, a volume, the denominator is the volume of the cylindrical pencil of radiation striking the target in $t = 1\,\mathrm{s}$:

$$\frac{U(v,T)}{ct\,A} = \frac{I(v,T)}{c} = \frac{U(v,T)}{V}. \qquad (3.3.3)$$

The energy density normal to the target $ds$ is

$$u(v,T) = \frac{U(v,T)}{V} = \frac{I(v,T)}{c}. \qquad (3.3.4)$$

To find the volume of all pencils of radiation that can strike the target in one second, we integrate over the solid angle of a sphere with the target at the center. This yields $4\pi$. The energy density within the chamber falling on the target from all angles in one second is $4\pi$ times the energy density of one pencil of radiation:

$$u(v,T) = \frac{4\pi}{c}I(v,T) \qquad (3.3.5)$$

with units of $\mathrm{J\,s^{-1}\,m^{-3}}$.

## 3.4 The Stefan-Boltzmann Law: Emissivity

As discovered empirically by Stefan in 1879, and derived about five years later by Boltzmann, the rate of heat transfer by radiation $dq/dt$ per unit area of the surface of a radiator is

$$\frac{dq}{dt} = e\sigma T^4 \qquad (3.4.1)$$

where $e$ is the emissivity. Emissivity varies between 0 (a perfect mirror) and 1 (a blackbody). A mirror is a very poor radiator and a black surface is a very good radiator. The darker a surface is, the better it is as a radiator (or absorber) of heat. A polished copper surface has $e \cong 0.3$ but the same copper radiator coated with carbon black has an emissivity that approaches 1. Although the term was proposed by Kirchhoff, who also stipulated its ideal properties, an actual blackbody cavity was not constructed for experimental use until about 1895. The blackbody is an evacuated box of the kind already discussed, with a very small hole by which one may observe the radiation that leaks out with suitable optical (or other) instruments. In practical terms, we use this principle when we make a rough estimate of the temperature of a very hot furnace (hot enough to emit visible radiation) by looking into a peephole in the furnace door. The interior glows red, orange, yellow, or white, according to how high the temperature is. Radiation is emitted over a broad spectrum which may or may not include visible light, depending on the temperature. In the most general terms, we write radiation energy density or intensity as a function of two variables, for example $u(v, T)$ or $I(v, T)$, but not more than two.

## 3.5 Stefan's Law

Given that electromagnetic radiation has energy, it is reasonable to suppose that it has momentum. Possessing momentum, it can exert pressure. Radiation pressure is not a theoretical construct. It is real and, though small, it can be measured under ordinary laboratory conditions. At the temperatures that obtain in the interior of stars, radiation pressure can be several million atmospheres.

As for the particles of an ideal gas, pressure is exerted on the walls of a container by the change in momentum upon collision with a wall (see figure 2.13.1). Unlike gas particles, however, radiation is not conserved in a closed system. The amount of matter in a closed chamber is constant, but radiation can be created or destroyed by raising or lowering the temperature. Thus we should expect any equation of state we write for radiation

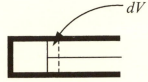

Figure 3.5.1. An isothermal cylinder fitted with a piston.

to be somewhat different from an equation of state for ideal gas particles. Gas particles have $pV = \frac{2}{3}\left(\frac{1}{2}mv^2\right) = \frac{2}{3}E_{kin}$ or, in terms of *kinetic energy density*, $u(T) = E_{kin}/V$,

$$p_{\text{ideal gas}} = \frac{2}{3}u(T). \tag{3.5.1}$$

By an analogous derivation (see equation 2.13.10), photons, which have two polarizations, exert a pressure of $2p_{rad} = 2u(T)/3$ or

$$p_{rad} = \frac{u(T)}{3} \tag{3.5.2}$$

(Huang 2001).

If radiation confined to a cylinder, fitted with a piston as in figure 3.5.1, undergoes a small isothermal, reversible expansion $dV$, the *entropy* increase $dS$ is, by the second law of thermodynamics,

$$dS = \frac{dq}{T}, \tag{3.5.3}$$

where $dq$ is the amount of heat that must be added to maintain thermal equilibrium during the expansion $dV$.

By the first law of thermodynamics, the energy increase of a system is

$$dU = dq + dw \tag{3.5.4}$$

where $dq$ and $dw$ are the heat and work going into the system; hence

$$dq = dU - dw = dU - p\,dV \tag{3.5.5}$$

for the pressure-volume work done on the system, and

$$dS = \frac{dq}{dT} = \frac{dU + p\,dV}{T} = \frac{1}{T}dU + \frac{p}{T}dV \tag{3.5.6}$$

where $pdV$ is the work done by the system *on the outside world* (which brings about a change in sign).

Energy, $U = f(V, T)$, is a state function; hence

$$dU = \left(\frac{\partial U}{\partial V}\right)_T dV + \left(\frac{\partial U}{\partial T}\right)_V dT. \tag{3.5.7}$$

Since $dS = (1/T) \, dU + (p/T) \, dV$, we have

$$dS = \frac{1}{T} \left[ \left( \frac{\partial U}{\partial V} \right)_T dV + \left( \frac{\partial U}{\partial T} \right)_V dT \right] + \frac{p}{T} \, dV$$

$$= \left[ \frac{1}{T} \left( \frac{\partial U}{\partial V} \right)_T + \frac{p}{T} \right] dV + \frac{1}{T} \left( \frac{\partial U}{\partial T} \right)_V dT. \qquad (3.5.8)$$

Entropy is also a state function, $S = f(V, T)$, so we have

$$dS = \left( \frac{\partial S}{\partial V} \right)_T dV + \left( \frac{\partial S}{\partial T} \right)_V dT. \qquad (3.5.9)$$

Equating the coefficients of $dV$ and $dT$,

$$\left( \frac{\partial S}{\partial V} \right)_T = \frac{1}{T} \left( \frac{\partial U}{\partial V} \right)_T + \frac{p}{T} \qquad (3.5.10)$$

and

$$\left( \frac{\partial S}{\partial T} \right)_V = \frac{1}{T} \left( \frac{\partial U}{\partial T} \right)_V. \qquad (3.5.11)$$

It is also true of a state function, for example entropy, that the order of differentiation is immaterial:

$$\frac{d^2 S}{dT dV} = \frac{d^2 S}{dV dT}, \qquad (3.5.12)$$

which leads to

$$\frac{\partial}{\partial T} \left[ \frac{1}{T} \left( \frac{\partial U}{\partial V} \right) + \frac{p}{T} \right]_V = \frac{\partial}{\partial V} \left[ \frac{1}{T} \left( \frac{\partial U}{\partial T} \right) \right]_T. \qquad (3.5.13)$$

The left-hand side is

$$\frac{\partial}{\partial T} \left[ \frac{1}{T} \left( \frac{\partial U}{\partial V} \right) + \frac{p}{T} \right]_V = -\frac{1}{T^2} \left( \frac{\partial U}{\partial V} \right) + \frac{1}{T} \left( \frac{\partial U}{\partial T \partial V} \right) + \frac{\partial}{\partial T} \frac{p}{T}$$

$$(3.5.14a)$$

and the right side is

$$\frac{\partial}{\partial V} \left[ \frac{1}{T} \left( \frac{\partial U}{\partial T} \right) \right]_T = \frac{1}{T} \left( \frac{\partial U}{\partial V \partial T} \right) + \underbrace{\left( \frac{\partial U}{\partial T} \right) \left( \frac{\partial \frac{1}{T}}{\partial V} \right)}_{0} \qquad (3.5.14b)$$

because $T$ and $V$ are independent variables in $U = f(V, T)$. The order of differentiation is immaterial, so the right side cancels the second term of the left side, leaving only

$$-\frac{1}{T^2} \left( \frac{\partial U}{\partial V} \right) + \frac{\partial}{\partial T} \frac{p}{T} = 0 \qquad (3.5.15)$$

or

$$\left(\frac{\partial U}{\partial V}\right)_T = T^2 \frac{\partial}{\partial T}\frac{p}{T}. \tag{3.5.16}$$

The *total* radiation density is a function only of $T$; therefore we do not need partial differential notation:

$$\frac{dU}{dV} = T^2 \frac{d}{dT}\frac{p}{T}. \tag{3.5.17}$$

We have, for the radiation density, $u(T) = U/V$ and $p = u(T)/3$, so $U = u(T)V$, $u(T) = 3p$, and

$$\frac{dU}{dV} = \frac{d}{dV}u(T)\,V = u(T)\frac{dV}{dV} = u(T) = 3p. \tag{3.5.18}$$

Also,

$$T^2 \frac{d}{dT}\frac{p}{T} = T^2 \left(\frac{T\frac{dp}{dT} - p\frac{dT}{dT}}{T^2}\right) = T\frac{dp}{dT} - p. \tag{3.5.19}$$

Combining the two sides of the equation,

$$3p = T\frac{dp}{dT} - p \tag{3.5.20}$$

or

$$4p = T\frac{dp}{dT}. \tag{3.5.21}$$

This simple separable differential equation

$$\frac{dp}{p} = 4\frac{dT}{T} \tag{3.5.22}$$

has the solutions

$$\ln p = 4\ln T + (\text{const}) \tag{3.5.23}$$

or

$$p = (\text{const}')T^4. \tag{3.5.24}$$

Written in terms of the energy density in joules per cubic meter,

$$u(T) = 3p = 3(\text{const}')T^4 = \beta T^4, \tag{3.5.25}$$

which is the *Stefan-Boltzmann radiation law* with the modern value of $\beta = 7.56 \times 10^{-16}\,\text{J m}^{-3}\,\text{K}^{-4}$ .

We have seen that $u(v, T) = (4\pi/c)I(v, T)$, that is, $I(v, T) = (c/4\pi)$ $u(v, T)$, or, equivalently, $I(\lambda, T) = (c/4\pi)u(\lambda, T)$. The *spectral concentration of the radiant excitance* is defined as

$$M_v = \frac{c}{4}u(v, T),\tag{3.5.26a}$$

such that $M_v = \pi I(v, T)$ or

$$M_\lambda = \frac{c}{4}u(\lambda, T)\tag{3.5.26b}$$

such that $M_\lambda = \pi I(\lambda, T)$.

One sometimes sees this law written for the *total* amount of radiation passing through unit area in unit time, called the *radiant excitance* $M(T)$,

$$M(T) = \frac{c}{4}\int_0^\infty u(\lambda, T)d\lambda = \sigma T^4,\tag{3.5.27}$$

where $c$ is the speed of electromagnetic radiation in a vacuum and $\sigma = 5.67 \times 10^{-8}$. The constants $\beta$ and $\sigma$ are reconciled with one another by writing

$$7.56 \times 10^{-16}\left(\frac{2.998 \times 10^8}{4}\right) = 5.67 \times 10^{-8}$$

with unit analysis

$$\mathrm{J\,m^{-3}\,K^{-4}}\left(\mathrm{m\,s^{-1}}\right) = \mathrm{J\,s^{-1}\,m^{-2}\,K^{-4}} = \mathrm{W\,m^{-2}\,K^{-4}},$$

where the watt (not a familiar unit for most chemists) is $\mathrm{W} = \mathrm{J\,s^{-1}}$.

Although we shall not rely on the terms $M(T)$ and $M_v$ in this book, they are mentioned here to clarify the origin of the well-known Stefan-Boltzmann constant $\sigma = 5.67 \times 10^{-8}\,\mathrm{W\,m^{-2}\,K^{-4}}$ and to emphasize the distinction between radiant energy over the entire blackbody spectrum and radiant energy over an infinitesimal part of it, the interval from $v$ to $v + dv$. Note the equivalence of the integral of energy density over wavelength and that over frequency:

$$\int_0^\infty u(\lambda)d\lambda = \int_0^\infty u(v)dv,\tag{3.5.28}$$

where $u(\lambda)$ and $u(v)$ are $u(\lambda, T)$ and $u(v, T)$ under the assumption that $T$ is constant. Frequency increases as wavelength decreases, so one is simply integrating forward or back under the same curve using a scaled horizontal axis. The area must be the same for both integrations.

We shall make several asides (section 3.7) concerning nomenclature, units, and notation, which can be inconsistent in this field and can lead to difficulties in reading the literature, especially when comparing early

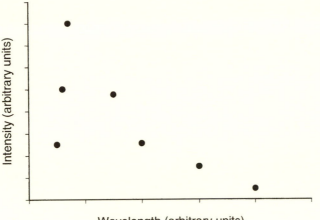

Figure 3.6.1. Intensity $I(\lambda, T)$ versus wavelength $\lambda$ for radiation from a copper radiator coated with soot (Langley 1886, see Kuhn 1987).

with late literature or textbooks written by chemists with those written by physicists.

## 3.6 The Blackbody Radiation Spectrum

To this point, while recognizing that radiation is a function of both temperature and wavelength, we have not addressed the spectral distribution of radiation, that is, the curve of $I(\lambda, T) = (c/4\pi)u(\lambda, T)$ versus $\lambda$. Accurate, controlled measurements on radiation intensity as a function of wavelength were made by the American astronomer S. P. Langley, who in 1886 published a remarkably prescient and complete study of radiation from a heated copper radiator coated with soot, a good approximation to a blackbody radiator, and, as it turned out, capable of producing a good replica of the sun's spectrum minus the Fraunhofer lines (which we won't go into here).

The few points in figure 3.6.1, taken from Langley's extensive and meticulous observations, serve to show that, whatever the functional relationship between radiation intensity $I(\lambda, T)$ and wavelength ($\lambda$) may be, it has a clear maximum; hence figures 3.6.2a and b show that $I(\lambda, T)$ cannot be a simple power or an inverse power of $\lambda$. This point will be important later.

The strength of Langley's work is that it gives a qualitative functional form to the relationship between intensity and wavelength. The weakness is that his intensity measurements were relative. For the theoretical studies

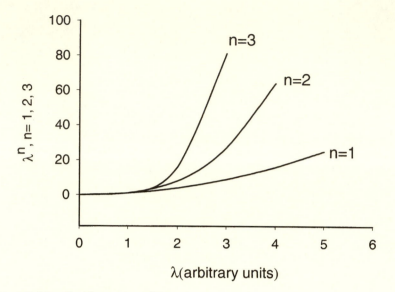

Figure 3.6.2a. Simple powers of $\lambda$.

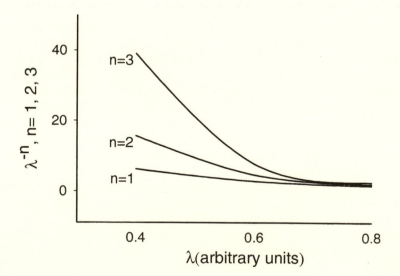

Figure 3.6.2b. Inverse powers of $\lambda$.

that had been started by Boltzmann and were soon to be pursued by, among others, Wien and Planck, absolute measurements were necessary.

## 3.7 Measurement of the Blackbody Spectrum

*Method*

If radiation falls on a target for $\Delta t$ seconds, and the temperature of the target rises by $\Delta T$ kelvins, the heat capacity of the target $C$ gives the total heat energy absorbed $q = C\Delta T$. The energy transferred to the target per unit time, $q/\Delta t$, is the power $P$. The electrical power is $P = i^2 R$ where $i$ is the current forced through a resistance $R$. If a detector circuit can be arranged as in figure 3.7.1 such that current can be induced or altered by incident radiation, then the power transferred to the detector from the radiator can be found by measuring the current through the known resistance of the detector.

Knowing the area of the aperture of the radiator, one can calculate the power per unit area, $q/\Delta t A = I(\lambda, T)$, where $I(\lambda, T)$ is the intensity. If the radiation at constant $T$ can be separated into its monochromatic components, one can carry out the measurement at numerous wavelengths so as to obtain the blackbody spectrum $I(\lambda)$ versus $\lambda$ comparable to figure 3.6.1 at any selected cavity temperature.

*Units*

The units for $\nu$ are reciprocal seconds, $s^{-1}$, that is, cycles per second, often called hertz, Hz. The unit for $\omega$ is the $\text{rad}\,s^{-1}$ where the radian (rad) is a segment of the circumference of a circle equal in length to its radius. The full circumference of a circle is $2\pi$ rad. Frequencies are expressed in different books as either $\omega$ or $\nu$ because the two units of measurement differ only by a constant, $\omega = 2\pi\nu$. The modern unit for $\lambda$ is often the nanometer, $\text{nm} = 10^{-9}$ m, although any rational subdivision of the meter is correct. The visible spectrum falls in the narrow range of about 400 to 700 nm. Somewhat longer wavelengths are heat radiation.

The unit of energy is the joule J and the unit of power is the watt $W = J\,s^{-1}$. The energy density $u$ is in $J\,m^{-3}$, the energy density per unit wavelength $u_\lambda$ is in $J\,m^{-4}$, and the energy density per unit frequency is in $J\,m^{-3}\,s$. The intensity $I$ has the units $J\,m^{-2}\,s^{-1} = W\,m^{-2}$, the intensity per unit wavelength has units of $W\,m^{-3}$, and $I$ per unit frequency has the units $J\,m^{-2}$. The spectral concentrations $M$ have the same units as intensity.

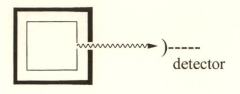

Figure 3.7.1. Schematic diagram of a blackbody radiator and a radiation detector.

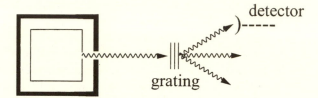

Figure 3.7.2. Schematic diagram of a blackbody radiator, diffraction grating, and detector.

*Apparatus*

To measure the blackbody spectrum, one needs a way of separating the wavelengths of radiation emitted by a blackbody into its monochromatic components. Nineteenth century physicists and analytical chemists were very good spectroscopists. The theory and practice of separating electromagnetic radiation by means of a diffraction grating (a transparent barrier scored with closely spaced lines) was well advanced by the last decade of the century (see figure 3.7.2). A diffraction grating serves the same function as a prism in separating radiation into its various wavelengths (frequencies), or *colors* if we choose to look at the special case of the visible portion of the electromagnetic spectrum. Radiation can either pass through the grating to be dispersed into its components as it is by a prism, or be reflected from a scored mirror to be dispersed into its component wavelengths. A detector set at a fixed angle to the grating receives only one wavelength. In principle at least, reflection gratings disperse radiation without otherwise acting upon it, that is, without absorbing any of it. Samuel Langley used a reflection grating.

A suitable detector produces some measurable signal in response to the intensity of radiation falling on it. The thermocouple and the thermopile (collection of thermocouples connected in series) were commonly used in the nineteenth century. A significant technical advance was made by Langley with his invention of the *bolometer*. Langley connected two fine blackened platinum filaments as opposite arms of a Wheatstone bridge such that radiation fell on one but not on the other. A sensitive galvanometer detected imbalance of the bridge brought about by the change in resistance in the illuminated filament relative to the dark filament. This enabled Langley to plot the curve shape of $I(\lambda)$ as a function of $\lambda$ of the illuminating radiation at the selected value of $T$. The bolometer is sensitive to radiation from about 600 nm to about 20 000 nm.

The rest of the experimental problem was simple but not easy. We do not know the proportionality constant between the intensity of radiation falling on the detector and the signal output (bridge voltage). We do not even know whether the detector collects all of the energy that falls on it, or whether it lets some slip by. It is not difficult to measure the power output $P$ of a detector relative to the power output of a standard (perhaps that of a "dark" detector $P_0$), but it is technically difficult to obtain the absolute intensity, which was needed for the future development of the theory of blackbody radiation.

Throughout the last decade of the nineteenth century, experimentalists grappled with these problems with results that were uniformly low because of power escape from the detector. Much work went into the problem of cutting down on power loss by several research groups, with results that gradually approached the curve that we now believe to be the correct one. One technical advance was to place the detector at the focus of a parabolic mirror so that radiation falling anywhere on the mirror would be reflected onto the detector (much like concentrating the rays of the sun with a magnifying glass). Another was the development, by the spectroscopist Paschen (for whom the Paschen lines in the hydrogen spectrum are named), of a remarkably sensitive D'Arsonval galvanometer-bolometer combination, which he claimed was capable of detecting temperature changes of $10^{-7}°$C (in modern units $10^{-7}$ K). As a culmination of this work, we now believe that the peak of the blackbody radiation curve at, for example, 2000 K falls at a wavelength of $\lambda_{max} = 1.449 \times 10^{-6}$ m, 1449 nm, and has an energy density per unit wavelength of $u(\lambda_{max}) = 5500$ J m$^{-4}$, which leads to the entire curve at 2000 K from the relative measurements that were already known. Knowing the curve at any temperature, one can, in principle, use the same detector to determine the curve at some other temperature, making measurements of the ratio of detector response at the new temperature relative to the corresponding response at the old temperature.

There are still many experimental problems, such as inability to obtain perfectly monochromatic radiation, variation of detector response with wavelength or temperature, stray radiation coming into the detector, scattering by the diffraction or reflection grating, and so on, but they are too technical to go into here. Suffice it to say that by the last decade of the nineteenth century, experimental physicists were able to achieve a high degree of success over the problems inherent in their craft (see

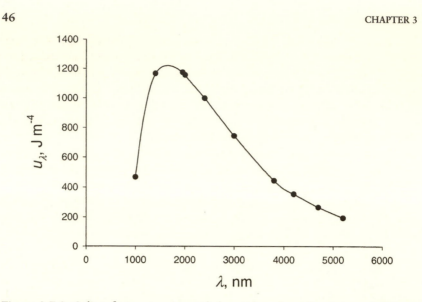

Figure 3.7.3. Spline fit to experimental data taken from Lummer and Prings-heim at 1643 K (1899, see Kittel 1980). A spline fit is an empirical curve-fitting procedure.

figure 3.7.3). Not only was gradual progress made in improving the quantitative measurement of the blackbody radiation curve, but the range of wavelengths over which reliable measurements could be made was also augmented. As we shall see in the next chapter, is it not only possible, in principle, to *measure* any absolute value of radiation intensity over a radiation spectrum, but W. Wien showed that one can calculate the blackbody radiation spectrum at any temperature provided one has the spectrum at some other temperature. Theoretical attempts to explain or derive this distinctive function *from first principles* were, however, uniformly unsuccessful.

## 3.8 Astrophysical Data from the COBE Satellite

As a final note on this difficult experimental problem, see comments by Mather et al. (1990) on the detection of cosmic background radiation. Cosmic background radiation is presumed to be the residuum of radiation that was in equilibrium with matter within the universe about 0.5 million years after the big bang. At that time, the temperature of the universe was about 3000 K, and interconversion between matter and radiation ceased as the universe expanded with consequent cooling. Since then, the universe has cooled by a factor of about 1000 because it has expanded

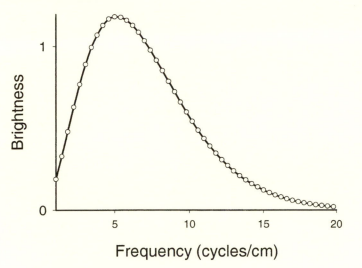

Figure 3.8.1. COBE cosmic microwave background radiation spectrum. The points are experimental measurements and the solid curve is the Planck equation. The units of brightness are $10^{-4}$ ergs $s^{-1}$ $cm^{-2}$ $sr^{-1}$ cm. (Adapted from Mather et al. 1990.)

(adiabatically of course) by a factor of about 1000. We are led to believe that intergalactic radiation from the big bang plus 0.5 million years will be at a temperature of about 3 K. The radiation should be *blackbody radiation* with the entire universe as the blackbody radiation chamber.

Radiation was found in this region as early as 1962, but its measurement was complicated by absorption by the earth's atmospheric gases. In 1989, a satellite, the COBE (cosmic background explorer), was sent up with the intention of measuring cosmic radiation above the earth's atmosphere. Preliminary results from the COBE satellite were reported in 1990 (figure 3.8.1). This paper contains a striking example of the blackbody radiation spectrum. The intergalactic temperature measured by COBE is $2.735 \pm 0.06$ K.

As an update on the COBE results, Mather reported to a scientific meeting in 1993 that the data of the COBE satellite have been analyzed more completely, reducing the uncertainty to 0.03% and the intergalactic temperature to $2.726 \pm 0.01$ K. Further updates are available on the web at **http:aether.lbl.gov**. See also Kittel and Kroemer 1980 (rev. 2003). The question of whether cosmic background radiation is perfectly isotropic or not has very interesting implications. Experimental work in this field promises to extend well into the twenty-first century.

## PROBLEMS

**3.1.** Show that $J\,m^{-4}$ and $W\,m^{-3}$ are acceptable units for $u_\lambda$ and $I_\lambda$ (section 3.7).

**3.2.** If the temperature implied by the cosmic background radiation curve is 2.73 K and if the Stefan-Boltzmann radiation law holds, what is the average density of radiation $u(T)$ in the universe? Give units.

**3.3.** The rate of heat emission by an object of emissivity $e$ is given by equation 3.4.1 per unit area or

$$\frac{dq}{dt} = e\sigma A\, T^4$$

for a radiator of area $A$. If a hot object at $T_2$ is placed in a thermostated chamber of $T_1$ where $T_2 < T_1$, what is the rate of heat loss from the object to the chamber?

**3.4.** The filament in a 100 W light bulb has a surface area of 40.0 mm². Its emissivity is 0.800. (a) Assuming perfectly efficient power consumption, what is its operating temperature? (b) What would its operating temperature be if it radiated as a perfect blackbody?

**3.5.** A blackbody cavity having a 10.0 mm² aperture is maintained at 1400 K. What is the power leak through the aperture?

**3.6.** A polished copper sphere of radius 1.00 mm, initially at 400 K, is suspended in an evacuated chamber maintained at 300 K. What is the initial heat loss from the sphere? (Recall that polished copper has an emissivity of about 0.30; see section 3.4.)

**3.7.** What is the heat loss from the sphere in problem 3.9.6 after a very long but finite time?

# Four

## The Planck Equation

IN 1893 W. WIEN PROPOSED on theoretical grounds that the intensity of blackbody radiation relative to its maximum intensity, $I_{\lambda T} : I_{\lambda_{\max} T}$, is the same function of $\lambda/\lambda_{\max}$ for all values of $T$, where $\lambda_{\max}$ is the wavelength of maximum intensity at any selected $T$. By 1896 Paschen provided extensive experimental corroboration of Wien's law of intensity ratios. In time, the theoretical foundations of this contention were to be discarded in favor of a much more radical theory, and Wien's general equation for energy density as a function of wavelength was to be relegated to the status of an approximate solution to the problem. Nevertheless, Wien's provisional theory and the ratio laws were the first significant steps toward a theory of blackbody radiation.

### 4.1 The Paschen-Wien Law

To see how the Paschen-Wien law works, consider the two experimental blackbody radiation spectra shown in figure 4.1.1, observed at different temperatures.

First, replace $\lambda$ on the horizontal axis with the ratio $\lambda/\lambda_{\max}$. This "normalizes" the two curves so that the peak of each falls at $\lambda/\lambda_{\max} = 1.0$. Now we attempt to "normalize" the heights of the two curves by plotting the ratio $u_{\lambda,T}/u_{\lambda_{\max},T}$ (which is equivalent to plotting $I_{\lambda T} : I_{\lambda_{\max} T}$) on the vertical axis (figure 4.1.2). Behold, it works! Both normalized curves are the same. The peak of both curves is at the ratio $u_{\lambda,T}/u_{\lambda_{\max},T} = 1.0$.

This is a very important discovery. By Wien's law, if we can find the distribution function for one temperature, we have the distribution function for all temperatures. As a consequence of the Paschen-Wien law, $\lambda_{\max} T = \text{const} \equiv b_W$. In 1897 Lummer and Pringsheim published an excellent value of $b_W = 0.294 \, \text{cm deg}$ (modern value, $b_W = 0.290 \, \text{cm K} = 2.898 \times 10^{-3} \, \text{m K}$). We shall see that this law can be derived from the distribution function of $u_\lambda$ versus $\lambda$ or, equivalently, $u_\nu$ versus $\nu$. We now meet the central problem of finding what that distribution function is. We wish to derive the universal blackbody radiation spectrum of figure 4.1.3.

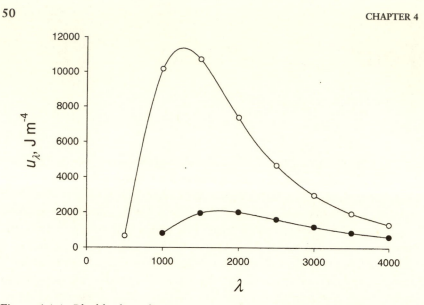

Figure 4.1.1. Blackbody radiation spectra of energy density as a function of wavelength. The upper curve is taken from statistically smoothed experimental points at $T = 2320\,\text{K}$ and the lower curve is at $1643\,\text{K}$.

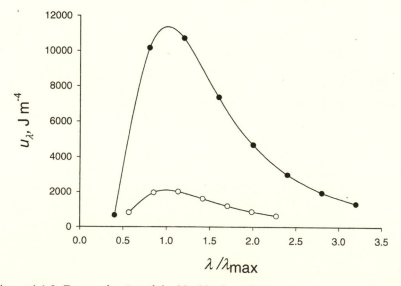

Figure 4.1.2. Energy density of the blackbody radiator as a function of the ratio $\lambda/\lambda_{max}$ at 2320 and 1643 K. The horizontal position of the two curves has been "normalized" so that the peaks fall at $\lambda/\lambda_{max} = 1.0$.

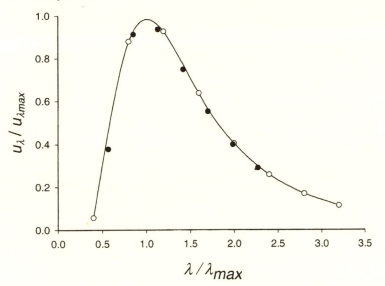

Figure 4.1.3. The ratio $u_\lambda/u_{\lambda_{max}}$ as a function of $\lambda/\lambda_{max}$. Normalization of both variables causes all points from the two temperatures to fall on the same curve. The maximum of the curve falls at the point $(1, 1)$. Points from any other temperature would fall on the curve as well.

## 4.2 Fitting the Curve

It is not difficult to fit a peaked curve like the blackbody radiation distribution to an analytical function. One needs a sharply rising function like $ax^2$ or $ax^3$ to fit the curve near the origin and a diminishing function like $e^{-bx}$ to fit the curve away from the origin. Between the rise of the first function and the fall of the second, there must be a peak if the product function is not to be zero everywhere (figure 4.2.1).

Another form is $y = x^{-n}e^{-1/x}$, shown in figure 4.2.2 for $n = 4.5, 5.0$, and $5.5$. Note that values of $x$ are $< 1.0$. Clearly, these curves "have the look" of the experimental curves in figure 4.1.1.

In 1894, Wein arrived at a preliminary empirical equation for the blackbody spectrum, finding

$$u_\lambda = a\lambda^\gamma e^{-f} \tag{4.2.1}$$

where $a$ is a constant, $\gamma$ is an exponent somewhere in the range of $-5$ to $-6$, and $f$ is a function as yet unknown. In this work, Wien was simply seeking an analytical equation that would express the functional behavior of $u_\lambda$ versus $\lambda$. He was not yet seeking a derivation from first principles.

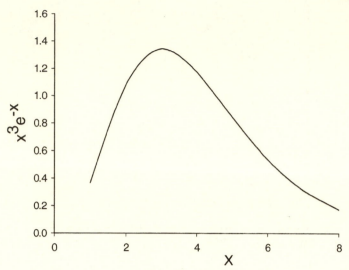

Figure 4.2.1. The function $y = x^3 e^{-x}$.

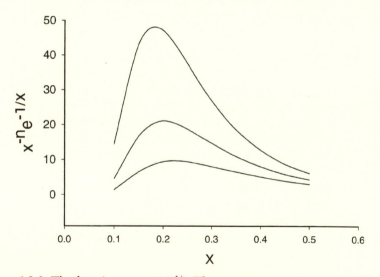

Figure 4.2.2. The function $y = x^{-n} e^{-1/x}$. The exponent $n$ is 4.5, 5.0, and 5.5 for the three curves.

In 1895 Paschen, in collaboration with Wien, proposed the form

$$u_\lambda = a\lambda^\gamma e^{-b/\lambda T}. \tag{4.2.2}$$

At this point it is convenient to express the energy density function in terms of frequency as was done by most theoreticians of the day. The conversion can be done by noting that

$$u_\lambda d\lambda = -u_v dv$$

where the minus sign appears because of the inverse relationship between $\lambda$ and $v$. (As $\lambda$ increases $v$ decreases.) It will drop out because, whether we integrate over $v$ from 0 to $\infty$ or over $\lambda$ from 0 to $\infty$, we are integrating the same experimental data but in opposite directions.

Recalling that $\lambda = c/v$ where $c$ is the speed of electromagnetic radiation in a vacuum,

$$u_v = -u_\lambda \frac{d\lambda}{dv} = -a\lambda^\gamma e^{-b/\lambda T} \frac{d\left(\frac{c}{v}\right)}{dv}$$

$$= -a\left(\frac{c}{v}\right)^\gamma e^{-bv/cT} c\left(\frac{-1}{v^2}\right) = ac^{\gamma+1}\left(\frac{1}{v}\right)^{\gamma+2} e^{-bv/cT}$$

$$= ac^{\gamma+1} v^{-(\gamma+2)} e^{-(b/c)/(v/T)}. \tag{4.2.3}$$

Now let $-(\gamma + 2) = n$, $ac^{\gamma+1} = \alpha$, and $b/c = \beta$, and we get

$$u_v = \alpha v^n e^{-\beta v/T}. \tag{4.2.4}$$

Wien soon showed that the exponent $n$ has to be 3 if the equation is to be consistent with Stefan's law, $u = \sigma T^4$, which was by that time widely accepted (see section 3.5, equation 3.5.26).

The total radiation density $u$ at all frequencies is

$$u = \int_0^\infty u_v dv = \alpha \int_0^\infty v^n e^{-\beta v/T} dv = \frac{\alpha\, n!}{\left(\frac{\beta}{T}\right)^{n+1}}, \tag{4.2.5}$$

where we have used the known definite integral

$$\int_0^\infty x^n e^{-ax} = \frac{n!}{a^{n+1}}.$$

If $n = 3$, then

$$u = \frac{\alpha\, n!}{\left(\frac{\beta}{T}\right)^{n+1}} = \frac{6\alpha}{\left(\frac{\beta}{T}\right)^4} = \frac{6\alpha}{\beta^4} T^4 = \text{const} \times T^4, \tag{4.2.6}$$

TABLE 4.2.1. Blackbody radiation energy
densities for two temperatures.

| $\lambda$, nm | $u_\lambda$, J m$^{-4}$ $T = 1646$ K | $u_\lambda$, J m$^{-4}$ $T = 2320$ K |
|---|---|---|
| 500.0 | — | 660.0 |
| 1000.0 | 810.0 | 10167.0 |
| 1500.0 | 1961.0 | 10720.0 |
| 2000.0 | 2013.0 | 7365.0 |
| 2500.0 | 1607.0 | 4675.0 |
| 3000.0 | 1185.0 | 2980.0 |
| 3500.0 | 856.0 | 1949.0 |
| 4000.0 | 620.0 | 1314.0 |

which is Stefan's law. Had we started with any other "rising function"
(figure 4.2.1), say $av^2$ or $av^{3.8}$, we would not have arrived at Stefan's law.

We can also find $\gamma$ for the second form of Wien's law, the one in terms
of $\lambda$. Since $n = 3 = -(\gamma + 2)$, one arrives at $\gamma = -5$ as the only exponent
in

$$u_\lambda = a\lambda^\gamma e^{-b/\lambda T}$$

that is compatible with Stefan's law. Now we have

$$u_v = \alpha v^3 e^{-\beta v/T} \tag{4.2.7a}$$

or

$$u_\lambda = c_1 \lambda^{-5} e^{-c_2/\lambda T} \tag{4.2.7b}$$

as two equivalent forms of what we now call the Wien equation. Either
of these equations is a good fit to the blackbody radiation distribution
function as it was known at the time.

As an important aside, we should mention that an equivalent way of
normalizing the blackbody spectrum for different temperatures is to plot
$u_\lambda/T^5$ versus $\lambda T$ at any fixed $T$. If this procedure is applied to a second
data set collected at a different temperature, the points will fall on the same
curve, the same is true for a third temperature, and so on. Applying these
transformations to the data set in table 4.2.1 and plotting the resultant
points yields figure 4.2.3.

The important feature is that, just as $\lambda_{max} T$ gives a single point on the
universal blackbody radiation curve, *any other $\lambda T$ gives another point
on the same curve*, that is, the entire curve is a function of the composite
variable $\lambda T$. The composite variable $\lambda T$ arises from the Paschen-Wien
displacement law, and the scale factor $u_\lambda/T^5$ arises from Stefan's law.

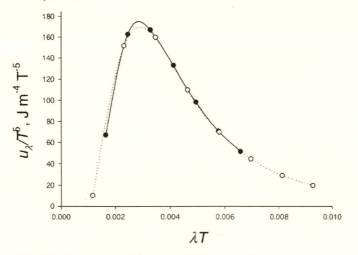

Figure 4.2.3. The function $u_\lambda/T^5$ versus $\lambda T$ for two fixed temperatures 1646 K and 2320 K. The vertical and horizontal axes are scaled.

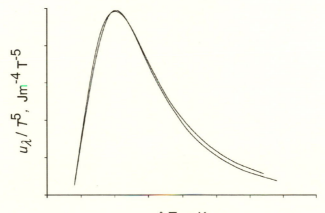

Figure 4.2.4. Wien's law and the experimental curve for blackbody radiation. The lower curve is from Wien's law. As the wavelength range is extended beyond $\lambda = 4000\,\text{nm}$, the systematic discrepancy between Wien's equation and experiment becomes more pronounced.

The Wien equation fitted the experimental curve so well that both Wien and the famous thermodynamicist Max Planck labored for two years to obtain a rigorous theoretical derivation of it. Both were unsuccessful. Moreover, new experimental work showed that the Wien equation is almost, but not quite, right. As the range of accurate measurements was extended into the infrared region of the spectrum, first to 6000 nm, then to 8000 nm, discrepancies, which might have been attributed to experimental error in 1896, were more and more disturbing. The turn of the twentieth century was a discouraging time. A rigorous theory of blackbody radiation was not forthcoming. What had seemed to be a solution to the problem was showing undeniable signs of failure.

## 4.3 The Number Density of Oscillatory Modes

The allowed frequencies of standing waves for a vibrating system are limited by the boundary conditions imposed upon the system. Each distinct wave is a *mode* of oscillation. Each allowed frequency is a *state* of the system. Two or more modes may have the same frequency, contributing to the same state. States having more than one mode are called *degenerate* states.

In dealing with blackbody radiation and the Planck-Einstein heat capacity theory, we shall want to know the energy density $u_v(v, T)$ in the interval of frequency $v$ to $v + dv$. In order to find this energy density for a cavity or a crystal of some simple geometry such as a cube, we need to know the number of oscillatory modes that can occur up to and including some selected frequency, on the presumption that each mode will contribute to the energy of the system in some way.

If we ask how many modes can exist without destructive interference for a vibrating string up to and including some specific frequency, say the third harmonic, the answer is three: the fundamental and the first and second overtones (figure 4.3.1). We can characterize these modes by their *wave numbers*, $m = 1$ for the fundamental and $m = 2$ and $m = 3$ for the first and second overtones. The third overtone ($m = 4$) has a wavelength that is shorter and hence a frequency that is greater than the specified limiting frequency. The point $m = 0$ is trivial (expresses the state of no vibration). The angular frequency $\omega$ of each mode is $\omega = m\pi v/L$ (equation 2.6.2) for a string of length $L$ where $v$ is the wave propagation speed. The total energy is given by the sum of contributions from the modes. For a vibrating string, the number of modes goes up as the limiting frequency

$$\omega = \frac{m\pi v}{L},$$

$$dm = \frac{L}{\pi v} d\omega. \tag{4.3.1}$$

The number density of modes in a linear wave number space increases as a linear function of increasing $m$. There is no degeneracy.

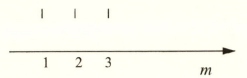

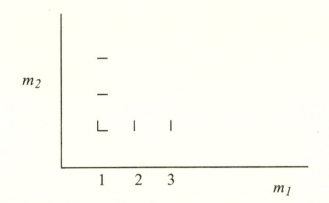

Figure 4.3.1. Oscillatory modes of a vibrating string in a one-dimensional wave number space. The limiting wave number is specified as 3 and there are three modes. There is no degeneracy.

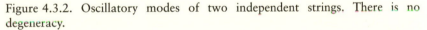

Figure 4.3.2. Oscillatory modes of two independent strings. There is no degeneracy.

Two vibrating strings perpendicular to one another as in figure 4.3.2 produce a total of six modes of vibration up to the third harmonic, three for each string vibrating independently of the other. If the strings are plucked with the same energy (amplitude), there is twice as much vibratory energy for both as for one.

If we place the same restriction on two-dimensional vibration, however, say a square vibrating membrane of length $L$ on an edge, we find that the number of modes is more than six. In the space spanned by the wave numbers $n_1$ and $n_2$, we have eight (see figure 4.3.3 [left]). The eight modes correspond to combinations of integers $(n_1, n_2)$ of $(1, 1), (1, 2), (2, 1), (2, 2), (1, 3), (3, 1), (2, 3),$ and $(3, 2)$, all of which have frequencies up to and including the third harmonic. If each mode were to contribute the same amount of energy, the oscillation would have more than twice the energy of a one-dimensional oscillator with all other parameters the same.

As a diagrammatic device, let us represent allowed modes by squares, such that each edge of the square connects points representing two allowed modes.

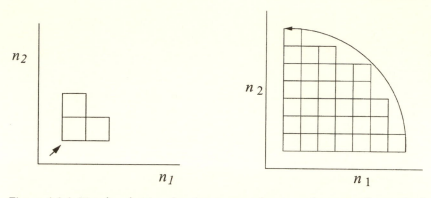

Figure 4.3.3. Number density plots in wave number space for a two-dimensional vibrating surface. The short arrow on the left points to the corner of the square representing the $(1, 1)$ solution. As the number of modes increases, the limiting area in wave number space approaches a quadrant.

The $(1, 1)$ mode (short arrow on the left of figure 4.3.3) is not degenerate. The $(1, 2)$ mode is degenerate with the $(2, 1)$ mode. The $(2, 2)$ mode is not degenerate; the $(1, 3)$ and $(3, 1)$ modes are degenerate, as are the $(2, 3)$ and $(3, 2)$ modes, making a total of eight modes within the limiting frequency. The number of vibrational modes goes up as $1, 2, 1, 2, 2$, etc., where 2 designates twofold degeneracy. This counting of modes is tedious and not easily generalized to many modes in three dimensions. We would like to have a well-behaved, general function describing the way the allowed number of modes increases with $n$. Let us call the number of modes $N$. We seek $N = f(n)$.

The corner of each square in figure 4.3.3 represents an allowed mode of oscillation. As the upper limit on $n_1$ and $n_2$ becomes larger, the pattern of squares representing allowed modes of oscillation of a vibrating membrane approaches the area of one-quarter of a circle of radius $n$ where, by Pythagoras's theorem, $n = (n_1^2 + n_2^2)^{1/2}$. The area of the entire circle is $A = \pi n^2$ but, because only positive values of $n_1$ and $n_2$ are physically meaningful, only the positive quadrant of the circle shown in figure 4.3.3 is meaningful. Each square represents a single mode because, although it has four corners, each corner is shared with four other squares so that, in the limit of large $n$, the number of corners (modes) is equal to the number of squares. The number of squares in the quadrant approaches $\frac{1}{4}\pi n^2$. Because $n_1$ and $n_2$ are integers, each edge has a length of one in wave number space.

For large values of $n_1$ and $n_2$, the angular frequency $\omega$ goes up as

$$\omega = \frac{v\pi}{L}\left(n_1^2 + n_2^2\right)^{1/2} = \frac{v\pi}{L}n \qquad (4.3.2)$$

as we saw in equation 2.9.9 for the two-dimensional wave equation. The magnitude of the radius vector of the quarter circular arc in two-dimensional wave number space is

$$n = \frac{\omega L}{v\pi} \tag{4.3.3}$$

and

$$\frac{1}{4}\pi n^2 = \frac{1}{4}\frac{\pi\omega^2 L^2}{v^2\pi^2} = \frac{\omega^2 L^2}{4\pi v^2} = \frac{\omega^2}{4\pi v^2}A \tag{4.3.4}$$

where $A$ is the limiting area of all the squares within distance $n$ from the $(1, 1)$ mode (essentially the center of the circle) for large $n$. The length of the vector $n$ limits the wave number, imposing a corresponding limit on the frequency. For large $n$, each area of a unit square represents a new allowed vibratory mode. The number of unit squares (modes) within a quadrant of radius $n$ is

$$N = \frac{1}{4}\pi n^2 = \frac{\omega^2}{4\pi v^2}A. \tag{4.3.5}$$

The *number density* of vibratory modes $\rho_N$ is the number of modes per unit area. Dividing the expression for $N$ by $A$,

$$\rho_N = \frac{N}{A} = \frac{\omega^2}{4\pi v^2}. \tag{4.3.6}$$

The number density in the range $\omega$ to $\omega + d\omega$ is

$$d\rho_N = \frac{2\omega}{4\pi v^2}(d\omega) = \frac{\omega}{2\pi v^2}d\omega. \tag{4.3.7}$$

Since $\omega = 2\pi v$, then $d\omega = 2\pi\, dv$ and

$$d\rho_N = \frac{2\pi v}{v^2}dv. \tag{4.3.8}$$

Thus the number density of modes increases linearly with $v$. Like a real vibrating string, a real vibrating thin sheet has an infinite number of harmonics which die out rapidly as the frequency increases. What one hears of a drumbeat is mainly the fundamental mode of vibration.

For the three-dimensional case, the pattern of allowed modes in wave number space becomes more complicated. Now we see the advantage of our detailed analysis of the number of modes of motion of a vibrating membrane. It is easily generalized to any wave number and to any number of dimensions. Figure 4.3.4 shows that, just as the number of modes in the two-dimensional case approached the area of a quadrant, allowed modes below a specified wavelength (a specified limit on $n$) in the three-dimensional case approach the volume of a spherical octant of radius $n$.

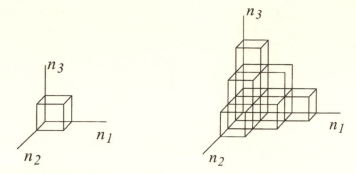

Figure 4.3.4. Density volumes for a three-dimensional vibration.

Generalizing from the two-dimensional case to the three-dimensional cube,

$$\omega = \frac{v\pi}{L}(n_1^2 + n_2^2 + n_3^2)^{\frac{1}{2}} = \frac{v\pi}{L}n,$$

$$N = \left(\frac{1}{8}\right)\frac{4}{3}\pi n^3$$

for the spherical octant, which leads to

$$d\rho_N = \frac{\omega^2}{2\pi^2 v^3}d\omega = \frac{4\pi v^2}{v^3}dv. \qquad (4.3.9)$$

For mechanical vibrations in a cubic three-dimensional wave number space, the number density of modes goes up as the square of the frequency. Other characteristics of a vibrating cubic solid (degeneracy, nodes, etc.) are analogous to what we have seen in one and two dimensions.

## 4.4 The Rayleigh-Jeans Equation

Suppose, applying the law of equipartition of energy, each mode of electromagnetic oscillation in an enclosed cavity at temperature $T$ has $k_B T$ of thermal energy. Combining $k_B T$ per vibrational mode with the number density of modes leads to the energy density as a function of frequency,

$$du_v = k_B T d\rho_N,$$

$$du_v = \frac{8\pi v^2}{c^3}k_B T dv, \qquad (4.4.1)$$

where we have doubled the number density of modes because of the twofold polarization (electric and magnetic) of electromagnetic radiation. Also we have replaced the speed $v$ with the speed $c$. This is the Rayleigh-Jeans equation. The Rayleigh-Jeans equation is a sound application of the

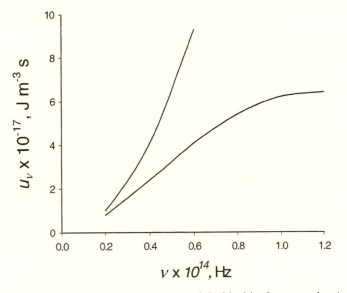

Figure 4.4.1. The Rayleigh-Jeans equation and the blackbody energy density curve calculated and observed at 2000 K. Experimental observations are shown by the lower curve.

classical principle of equipartition of energy to electromagnetic radiation within a chamber, but it leads to a completely unacceptable conclusion. If the number of allowed modes of oscillation within a blackbody radiator goes up as the square of the frequency, then the energy and the emitted intensity should also go up as the square of the frequency. If this conclusion were correct, the energy within any chamber at $T > 0$ K would be infinite because the term in $v^2$ has no upper limit. We know that when we look into an ordinary furnace we are not bombarded by X-rays and cosmic rays of near-infinite energy. We detect radiation mainly in the infrared (heat) with, perhaps, some radiation in the visible range. There must be some damping factor that refuses to allow emission of higher frequencies. We see damping of the Rayleigh-Jeans curve in the experimental behavior shown in the lower curve in figure 4.4.1.

The Rayleigh-Jeans equation, like the Wien equation, offers a tantalizing glimpse of a true theory for blackbody radiation. Its strength is that, unlike the Wien equation, it is rigorously derived from classical physical principles, but its weakness is, quite obviously, that it does not follow the experimental curve as closely as the Wien equation. It best approaches the experimental curve at low frequencies (long wavelengths), which is precisely where the Wien equation fails. Neither of these equations can be simply wrong; both tell us part of the blackbody radiation story. Prior to the Rayleigh-Jeans (1905) derivation, Max Planck had proposed a remarkable theory that was later shown by Einstein to reconcile the rigorous approach of Rayleigh and the semiempirical approach of Paschen and Wien to provide a complete theory of blackbody radiation. It will be seen that, at the extremities of frequency, the Planck equation reduces to the Rayleigh equation on the one hand and to the Wien equation on the other.

## 4.5 The Planck Equation

Planck's original derivation (1900) was founded on classical thermodynamics, but it used what was originally thought to be a mathematical artifact: treating energy as though it is discontinuous or *atomistic*. Since then equivalent but simpler derivations have been found, one of which, due to Einstein, we shall use here. All use the idea of discontinuous energy in one way or another.

Suppose that real or virtual oscillators (atoms perhaps) in the walls of an evacuated chamber (blackbody) are in thermal equilibrium with electromagnetic radiation contained in the chamber. Suppose further that the oscillators are divided into groups containing $n_0, n_1, n_2, \ldots$ oscillators having $0, 1, 2, 3, \ldots$ *units of energy*, each unit being proportional to frequency, $\varepsilon \propto v$, through a proportionality constant $h$

$$\varepsilon = hv. \tag{4.5.1}$$

These units were later called *quanta*.

This is the critical concept: At higher frequencies, the quanta are larger than they are at lower frequencies. At any given temperature there is a specific amount of thermal energy, $k_B T$, that is, a specific number of quanta, to be shared among the oscillators. There is a distribution of quanta of different frequencies within the chamber. At fixed temperature (energy), if we sample the intensity emitted by a blackbody radiator in a region of low frequency, we sample in a region of many small quanta. If we sample the intensity at high frequency, we are in a region at which only a few large quanta contribute to the total energy of the system at that frequency.

We have opposing trends. At low frequency, there are few modes in a small interval $dv$ due to the Rayleigh equation. They are satisfied or nearly satisfied by many quanta. Rayleigh's law holds or is approximated,

$$du_v = \frac{8\pi v^2}{c^3} k_B T dv. \tag{4.5.2}$$

At high frequencies, there are many modes in the interval $dv$ but not enough thermal energy to satisfy them all. There are not enough large quanta to go around, so some of the modes acquire a quantum of energy and some do not. The energy density of radiation increases by the increase in total number of modes times the *fraction* of modes that get a quantum of energy relative to total modes:

$$\frac{\text{modes acquiring a quantum of energy in } \delta v}{\text{total modes in } \delta v}.$$

The fraction of the modes getting a quantum relative to the total number of modes is equal to or less than 1; hence

$$du_v \leq \frac{8\pi v^2}{c^3} k_B T dv \qquad (4.5.3)$$

The curve of energy density versus frequency falls below the Rayleigh curve as in figure 4.4.1. In the limit of very high frequency, the fraction of oscillators acquiring a quantum of energy upon increasing $v$ to $v + dv$ becomes very small; hence the energy density of radiation within the chamber at very high frequency approaches zero. A nonzero, non-negative curve that approaches zero at both ends must have a maximum in the middle as in figure 4.1.1.

Rayleigh and Jeans were correct in recognizing the importance of the number of modes of motion, and Wien correctly guessed that the fraction of oscillators contributing to the energy density within a blackbody resonator is not 1 but is less than 1 as governed by a diminishing distribution. It was Planck who perceived that energy, like matter, is discontinuous in nature, the concept that reconciled the Rayleigh and the Paschen-Wien laws into a single successful theory governing blackbody radiation and, as it turns out, everything else in the physical universe.

Let us continue by applying the Boltzmann distribution (section 2.11) to groups of oscillators $n_0, n_1, n_2, \ldots$ having $0, 1, 2, 3, \ldots$ units of energy:

$$n_1 = n_0 e^{-(\varepsilon_1 - \varepsilon_0)/k_B T},$$

$$n_2 = n_0 e^{-(\varepsilon_2 - \varepsilon_0)/k_B T},$$

$$n_3 = n_0 e^{-(\varepsilon_3 - \varepsilon_0)/k_B T},$$

$$\vdots \qquad . \qquad (4.5.4)$$

Arbitrarily setting $\varepsilon_0 = 0$ and applying the quantum hypothesis $\varepsilon = hv$,

$$n_1 = n_0 e^{-\varepsilon_1/k_B T} = n_0 e^{-hv/k_B T},$$

$$n_2 = n_0 e^{-\varepsilon_2/k_B T} = n_0 e^{-2hv/k_B T},$$

$$n_3 = n_0 e^{-\varepsilon_3/k_B T} = n_0 e^{-3hv/k_B T},$$

$$\vdots \qquad (4.5.5)$$

provided that the energy levels are equally spaced, as they are for *harmonic* oscillators. The total number of oscillators, $n$, is

$$n = n_0 + n_1 + n_2 + \cdots$$

$$= n_0 + n_0 e^{-h\nu/k_B T} + n_0 e^{-2h\nu/k_B T} + n_0 e^{-3h\nu/k_B T} + \cdots$$

$$= n_0 \left( 1 + e^{-h\nu/k_B T} + e^{-2h\nu/k_B T} + e^{-3h\nu/k_B T} + \cdots \right). \qquad (4.5.6)$$

Let

$$e^{-h\nu/k_B T} = x,$$

$$e^{-2h\nu/k_B T} = e^{-h\nu/k_B T} e^{-h\nu/k_B T} = x^2,$$

$$e^{-3h\nu/k_B T} = e^{-h\nu/k_B T} e^{-h\nu/k_B T} e^{-h\nu/k_B T} = x^3,$$

$$\vdots \qquad\qquad (4.5.7)$$

so that

$$n = n_0 \left( 1 + x + x^2 + x^3 + \cdots \right). \qquad (4.5.8)$$

Now, using an old mathematician's trick, take the sum in parentheses,

$$1 + x + x^2 + x^3 + \cdots = \text{sum}, \qquad (4.5.9)$$

multiply by $x$,

$$x + x^2 + x^3 + \cdots = x\,(\text{sum}), \qquad (4.5.10)$$

and subtract the second equation from the first,

$$\text{sum} - x\,(\text{sum}) = 1,$$

$$\text{sum}\,(1 - x) = 1,$$

$$\text{sum} = \frac{1}{(1 - x)}. \qquad (4.5.11)$$

This gives us a closed value for our infinite sum which enables us to write a closed expression for $n$,

$$n = n_0 \left( 1 + x + x^2 + x^3 + \cdots \right),$$

$$n = n_0\,(\text{sum}) = n_0 \left( \frac{1}{1 - x} \right) = n_0 \left( \frac{1}{1 - e^{-h\nu/k_B T}} \right), \qquad (4.5.12)$$

and for $n_0$,

$$n_0 = n\left( 1 - e^{-h\nu/k_B T} \right). \qquad (4.5.13)$$

The total energy is

$$E = n_0(0) + n_1(hv) + n_2(2hv) + n_3(3hv) + \cdots$$
$$= 0 + n_0 hv e^{-hv/k_B T} + n_0 2hv e^{-2hv/k_B T} + n_0 3hv e^{-3hv/k_B T} + \cdots$$

from equation 4.5.6, so that the above equation becomes

$$E = n_0 hv e^{-hv/k_B T} \left(1 + 2e^{-hv/k_B T} + 3e^{-2hv/k_B T} + \cdots\right). \qquad (4.5.14)$$

Again, let

$$e^{-hv/k_B T} = x$$

as in equation 4.5.7, and we get a new sum

$$E = n_0 hv\, x \left(1 + 2x + 3x^2 + \cdots\right). \qquad (4.5.15)$$

Using the same old mathematician's trick on the sum in parentheses,

$$s = 1 + 2x + 3x^2 + 4x^3 + \cdots,$$
$$xs = x + 2x^2 + 3x^3 + \cdots. \qquad (4.5.16)$$

Subtract the second sum from the first,

$$1 + 2x + 3x^2 + 4x^3 + \cdots = s$$
$$-[x + 2x^2 + 3x^3 + \cdots] = sx \qquad (4.5.17)$$

and we get the sum we just solved,

$$1 + x + x^2 + x^3 + \cdots = \text{sum} = \frac{1}{1-x}. \qquad (4.5.18)$$

It follows that

$$s - sx = \frac{1}{1-x},$$
$$s(1-x) = \frac{1}{1-x},$$
$$s = \frac{1}{(1-x)^2}, \qquad (4.5.19)$$

which gives us the energy in closed form:

$$E = n_0 hv\, x \left(\frac{1}{(1-x)^2}\right) = \left(\frac{n_0 hv\, e^{-hv/k_B T}}{(1 - e^{-hv/k_B T})^2}\right). \qquad (4.5.20)$$

We already know that

$$n_0 = n\left(1 - e^{-h\nu/k_B T}\right),$$

so

$$E = \left(\frac{n_0 h\nu\, e^{-h\nu/k_B T}}{\left(1 - e^{-h\nu/k_B T}\right)^2}\right) = \left(\frac{n\left(1 - e^{-h\nu/k_B T}\right) h\nu\, e^{-h\nu/k_B T}}{\left(1 - e^{-h\nu/k_B T}\right)^2}\right)$$

$$= \left(\frac{n h\nu\, e^{-h\nu/k_B T}}{\left(1 - e^{-h\nu/k_B T}\right)}\right). \tag{4.5.21}$$

If we divide the numerator and denominator by $e^{-h\nu/k_B T}$,

$$E = \frac{n h\nu}{\left(\dfrac{1}{e^{-h\nu/k_B T}} - 1\right)} = \frac{n h\nu}{\left(e^{h\nu/k_B T} - 1\right)}. \tag{4.5.22}$$

The average energy over all the oscillators is

$$\overline{E} = \frac{E}{n} = \frac{h\nu}{\left(e^{h\nu/k_B T} - 1\right)}. \tag{4.5.23}$$

Multiplying the average energy by the number density of oscillator modes gives the energy density

$$du_\nu(\nu, T) = \frac{8\pi\nu^2}{c^3}\overline{E} = \frac{8\pi h\nu^3}{c^3}\frac{1}{\left(e^{h\nu/k_B T} - 1\right)}d\nu \tag{4.5.24}$$

or

$$u_\nu(\nu, T) = \frac{8\pi\nu^2}{c^3}\frac{h\nu}{\left(e^{h\nu/k_B T} - 1\right)}, \tag{4.5.25}$$

which is the celebrated Planck equation for the energy density within a blackbody radiation cavity. (This derivation deviates from Planck's original work [Planck 1900] for the purpose of simplicity).

The Planck equation is written in several different forms, among them,

$$u_\nu(\nu, T) = 8\pi h\left(\frac{\nu^3}{c^3}\right)\frac{1}{\left(e^{h\nu/k_B T} - 1\right)} \tag{4.5.26}$$

or

$$d\overline{E}_\omega = \frac{V\,\hbar\omega^3}{\pi^2 c^3}\frac{d\omega}{\left(e^{\hbar\omega/k_B T} - 1\right)}, \tag{4.5.27}$$

where $\hbar = h/2\pi$ and $V$ is the volume of the blackbody, or

$$\rho_\nu(\nu, T)d\nu = \frac{8\pi h\nu^3}{c^3} \frac{d\nu}{(\exp(h\nu/k_B T) - 1)} \tag{4.5.28}$$

where $\exp(x) = e^x$, and $\rho_\nu$ is a common alternative notation for the energy density. In terms of wavelength,

$$u(\lambda, T) = \frac{8\pi hc}{\lambda^5} \frac{1}{(e^{hc/\lambda k_B T} - 1)}. \tag{4.5.29}$$

## 4.6 Immediate Deductions from Planck's Law

Just prior to his presentation of the radiation law in its finished form, Planck had published an empirical equation

$$u_\nu(\nu, T) = \frac{A\nu^3}{(e^{B\nu/T} - 1)} \tag{4.6.1}$$

as the last stepping stone to his great theoretical triumph. The experimentalists Rubens and Lummer and Pringsheim quickly verified the close fit between Planck's empirical form and experimental data. Comparing the theoretical and empirical forms,

$$u_\nu(\nu, T) = \frac{8\pi\nu^2}{c^3} \frac{h\nu}{(e^{h\nu/k_B T} - 1)}$$

and

$$u_\nu(\nu, T) = \frac{A\nu^3}{(e^{B\nu/T} - 1)},$$

makes it evident that $A = 8\pi h/c^3$ and $B = h/k_B$. Thus from the parameter $A$, extracted from the experimental data, and the speed of electromagnetic radiation, $c$, which was well known at the time, Planck obtained the numerical value for the universal constant $h$ that we now call Planck's constant. This value evolved somewhat as more accurate and extensive experimental data were gathered, but it was quite accurate even in the beginning. Data available in 1899 led to $6.885 \times 10^{-27}$ erg s. By 1900, it was fitted as $6.55 \times 10^{-27}$ erg s. The modern value is $6.626 \times 10^{-27}$ erg s (modern units $6.626 \times 10^{-34}$ J s).

Once having the value of $h$, and the empirical curve fit for $B$, $B = h/k_B$ was solved for $k_B$. Planck's value of $1.346 \times 10^{-16}$ erg deg$^{-1}$ was much superior to previous estimates. The currently accepted value is $1.381 \times 10^{-16}$ erg deg$^{-1}$ = $1.381 \times 10^{-23}$ J K$^{-1}$. The Avogadro number $N_A$ was not well known at the turn of the twentieth century but the universal gas constant was. Planck, having the Boltzmann constant $k_B$, which is the gas constant per particle, found the Avogadro number from $N_A = R/k_B$,

$$N_A = \frac{R}{k_B} = \frac{8.3144}{1.381 \times 10^{-23}} = 6.021 \times 10^{23}$$

where modern values are used for $R/k_B$ and the units for $R$ are $J\,K^{-1}mol^{-1}$.

Planck obtained $6.175 \times 10^{23}$ particles $mol^{-1}$ for Avogadro's number using data available to him at the time. This permits accurate calculation of atomic and molecular masses from relative atomic weights. For example, taking the weight of a mole of carbon to be $0.012\,kg$ ($12.00\,g$), one has the weight of a carbon atom as

$$\frac{0.01200}{N_A} = 1.993 \times 10^{-26}\,kg.$$

As a final achievement, Planck, knowing the faraday $F$ to be the amount of electrical charge on one mole of singly charged ions, calculated the unit of electrical charge, that is, the charge on the electron, as $4.69 \times 10^{10}$ esu $= 1.56 \times 10^{-19}$ coulombs $mol^{-1}$. The modern value is $e = F/N_A = 1.602 \times 10^{-19}$ coulombs $mol^{-1}$.

## PROBLEMS

4.1. Why does the normalization in section 4.1 lead to Wien's law $\lambda_{max}T = $ const?

4.2. Lummer and Pringsheim published numerous curves comparable to figure 4.1.1 in the years preceding the turn of the twentieth century. Simple visual location of $\lambda_{max}$ at five different temperatures gave the data in table 4.1.

TABLE 4.1. Experimental values of $\lambda_{max}$ at specified temperatures

| $\lambda_{max}$ nm | 1200 | 1700 | 2050 | 2350 | 3050 |
|---|---|---|---|---|---|
| $T$ K | 2320 | 1646 | 1449 | 1259 | 998 |

Calculate the Wien displacement constant for each observation and calculate the mean $k_W$. How does this compare with the modern value of $2.898 \times 10^{-3}$ m K? Is there a trend in the data?

**4.3.** What is the wavelength $\lambda_{max}$ in nanometers and the frequency $\nu_{max}$ in hertz and radians per second at maximum emission intensity of a blackbody maintained at 5000 K?

**4.4.** What is the rate of emission of energy (power) for a $10.0\,cm^2$ blackbody radiator at 5000 K? The Stefan-Boltzmann constant is $5.67 \times 10^{-8}\,J\,m^{-2}s^{-1}K^4$ (section 3.5).

**4.5.** Derive Wien's displacement law from Planck's energy density law, equation 4.5.29. Recall that at the maximum of a function, its first derivative with respect to the independent variable is zero. You will have to solve a transcendental equation by trial and error or by an iterative method to obtain the final result.

**4.6.** As part of the solution to problem 4.5 you had to solve the equation

$$5\left(e^x - 1\right) = xe^x$$

for $x$. This type of equation is not solvable by conventional means but can be solved by searching for the value of $x$ that makes $a = 5(e^x - 1) - xe^x$ as close to zero as possible. Write a simple computer program to do this. As a starting point, notice that arbitrarily setting $x = 1.0000$ yields a value of $a$ that is considerably larger than zero but that by increasing $x$ the function goes through a maximum and has a root between 4.9 and 5.

**4.7.** Show that the collection of constants in

$$\frac{hc}{4.966k_B} = T\lambda_{max}$$

does indeed yield

$$2.898 \times 10^{-3} = T\lambda_{max}.$$

**4.8.** Find the wavelength and the frequency (in Hz and rad s$^{-1}$) at maximum emission of blackbody radiation at 2000 K.

**4.9.** Find the energy density of a blackbody radiator at 2000 K and $2.00 \times 10^{14}$ Hz according to the Rayleigh-Jeans equation.

**4.10.** Find the energy density of a blackbody radiator at 2000 K and $2.00 \times 10^{14}$ Hz according to the Planck equation.

**4.11.** Show that the Stefan-Boltzmann law can be derived from Planck's law, equation 4.5.26. To do this you will need to use the standard integral

$$\int_0^\infty \frac{x^3}{e^x - 1}\,dx = \frac{\pi}{15}.$$

# Five

## The Einstein Equation

THE HEAT CAPACITY of a substance is, as the name indicates, a measure of its ability to absorb heat. A substance whose temperature rises over a large interval upon application of a given amount of heat has a low capacity to absorb heat, and a substance whose temperature rises only a little with the application of the same amount of heat has a large capacity to absorb heat. Mathematically, $C = \partial U / \partial T$, where $U$ is the amount of heat energy added. By the late nineteenth century, low-temperature heat capacities were being measured and discrepancies were found between experimental results and the simple law of Dulong and Petit stating that $C \cong 25 \, \text{J K}^{-1} \text{mol}^{-1}$. The most serious discrepancies were found for diamond which, as early as 1872, was found to have a molar heat capacity of only $3.2 \, \text{J K}^{-1} \text{mol}^{-1}$ at 223 K. It was these anomalous results for diamond that first attracted Einstein in 1907 to the general problem of heat capacities.

## 5.1 The Einstein Model

Regard a solid crystal as a three-dimensional lattice of independent isotropic harmonic oscillators, each atom being a mass tethered to its lattice point by Hooke's law forces brought about by the presence of all the other atoms. The forces can be expressed as components in three Cartesian coordinates, that is, there are three degrees of freedom for the kinetic energy and three degrees of freedom for the potential energy. By our rule that each degree of freedom contributes $\frac{1}{2} k_B T$ to the energy and $\frac{1}{2} k_B$ to the heat capacity, we would expect that $C = 6 \frac{1}{2} k_B N_A = 3R \cong 25 \, \text{J K}^{-1} \text{mol}^{-1}$. This calculation is certainly not true for diamond, so the rule $C = \frac{1}{2} k_B$ per atom per degree of freedom must fail in this case. Einstein postulated that the system of crystal oscillators is quantized just as the blackbody radiator is, and that it follows Planck's law for the average energy $\overline{E}$, which limits its heat capacity in some circumstances.

## 5.2 Einstein's First Derivation: The Heat Capacity of Diamond

In his 1907 paper, Einstein first presented an alternative (simpler) derivation of the Planck blackbody radiation equation using essentially the logic
we followed in section 4.5. Having completed this task, he turned his
attention to the heat capacity of diamond, and so to the problem of heat
capacities in general.

If the average energy over all oscillators is (from equation 4.5.23)

$$\overline{E} = \frac{E}{n} = \frac{hv}{\left(e^{hv/k_B T} - 1\right)} \tag{5.2.1}$$

as in the Planck derivation, then a crystalline solid, which can be regarded
as an Avogadro number of isotropic harmonic oscillators per mole, should
have a molar energy

$$U_E = 3N_A \overline{E} = 3N_A \frac{hv}{\left(e^{hv/k_B T} - 1\right)}, \tag{5.2.2}$$

where the subscript $E$ designates a physical quantity according to the
Einstein model. The multiplier 3 arises because isotropic oscillators have
three degrees of freedom, $x$, $y$, and $z$. We seek the heat capacity $C_E = dU_E/dT$. Multiplying and dividing by $k_B$, one can write $U_E$ as

$$U_E = 3N_A k_B \frac{hv}{k_B} \left(e^{hv/k_B T} - 1\right)^{-1} = \frac{3Rhv}{k_B} \left(e^{hv/k_B T} - 1\right)^{-1}, \tag{5.2.3}$$

where $N_A k_B = R$, which makes differentiation with respect to $T$ easy:

$$C_E = \frac{dU_E}{dT} = \frac{3Rhv}{k_B} \left(e^{hv/k_B T} - 1\right)^{-2} e^{hv/k_B T} \frac{hv}{k_B} T^{-2}. \tag{5.2.4}$$

(Differentiation is with respect to $T$ only because $v$ is a parameter of the
system, fixed by the quantum of energy unique to that system.)

The heat capacity of a solid is, according to Einstein,

$$C_E = \frac{3R \left(\frac{hv}{k_B T}\right)^2 e^{hv/k_B T}}{\left(e^{hv/k_B T} - 1\right)^2}. \tag{5.2.5}$$

Figure 5.2.1 shows that Einstein's heat capacity equation $C = f(T)$ fits
the experimental points for diamond very well. Further experimentation
produced a unique Einstein curve for each of many solids with good
agreement between experiment and theory.

In the limit of high temperature, $k_B T \gg hv$, and we expect many quanta
of thermal energy to be equally distributed over oscillators for a small

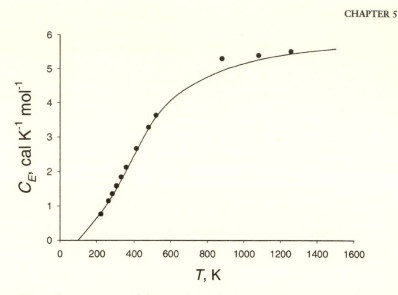

Figure 5.2.1. Heat capacity of diamond as a function of temperature. The solid line is the Einstein equation and the points are experimental data of H. Weber quoted by Einstein (1907). (In early experimental work, energy was reported in units of cal = 4.184 J.)

temperature increment. This is the Rayleigh-Jeans condition, so $\overline{E} = k_B T$. The condition $k_B T \gg h\nu$ leads to

$$C_E \to 3R \cong 25 \text{ J K}^{-1} \text{mol}^{-1} \qquad \text{(high temperature)},$$

which is, of course, the law of Dulong and Petit. What we mean by "high temperature" depends upon $h\nu$, which, in turn, depends upon the Hooke's law forces tethering atoms to their lattice sites within the crystal. For most metals, these forces are rather weak, $k_B T \gg h\nu$ at room temperature, and the law of Dulong and Petit holds. For some solids, especially diamond, the interatomic forces are strong and the law fails at room temperature. To observe Dulong and Petit behavior in diamond, one would need to go to temperatures of more than 2000 K.

At low temperature, the opposite condition is true, $k_B T \ll h\nu$, there are few quanta for a given small temperature increment, and $h\nu/k_B T$ is large. The difference $\left(e^{h\nu/k_B T} - 1\right)$ becomes essentially $e^{h\nu/k_B T}$; hence

$$C_E = \frac{3R \left(\frac{h\nu}{k_B T}\right)^2 e^{h\nu/k_B T}}{\left(e^{h\nu/k_B T} - 1\right)^2} \cong \frac{3R \left(\frac{h\nu}{k_B T}\right)^2 e^{h\nu/k_B T}}{\left(e^{h\nu/k_B T}\right)^2} = 3R \left(\frac{h\nu}{k_B T}\right)^2 e^{-h\nu/k_B T}.$$

$$(5.2.6)$$

As $T$ approaches $0\,\mathrm{K}$, $e^{-hv/k_BT}$, approaches $e$ to a very large negative exponent, that is, it also approaches zero. The term $e^{-hv/k_BT}$ approaches zero faster than $hv/k_BT$ increases, so

$$C_E \to 0 \qquad \text{(near } 0\,\mathrm{K}\text{)}.$$

## 5.3 The Einstein Temperature

Einstein's full $C_E$ equation has only one unknown parameter, $v$, which is unique to each solid. We can normalize the heat capacity curves for many solids by finding the temperature at which $k_BT = hv$ for each. This temperature is called the *Einstein temperature* and designated $\Theta_E$. Once having fixed the Einstein temperature, the *Einstein frequency* $v_E$ is also known:

$$k_B\Theta_E = hv_E. \qquad (5.3.1)$$

For many metals, $\Theta_E$ is around $200\,\mathrm{K}$.

If we plot the heat capacity $C$ for different solids, we observe a unique curve for each, as shown in figure 5.3.1a. (We will not distinguish between heat capacity at constant volume, $C_V$, and heat capacity at constant pressure, $C_P$, which are not very different for solids.) The curve for diamond extends to such high temperatures that only part of it appears in figure 5.3.1a. However, we also have a single point, $\Theta_E$, on each curve. If we plot $C$ as a function of the unitless ratio $T/\Theta_E$, we should be able to normalize the curves just as we did by plotting $\lambda/\lambda_{\max}$ in the Wien derivation. Figure 5.3.1b shows the *single* curve for all three solids, silver, aluminum, and diamond, found by plotting $C$ versus $T/\Theta_E$.

The data obviously fit the normalized Einstein equation

$$C_E = \frac{3R\left(\dfrac{hv}{k_BT}\right)^2 e^{hv/k_BT}}{\left(e^{hv/k_BT}-1\right)^2} = \frac{3R\left(\dfrac{\Theta_E}{T}\right)^2 e^{\Theta_E/T}}{\left(e^{\Theta_E/T}-1\right)^2} \qquad (5.3.2)$$

where $hv/k_B\Theta_E = 1$ and $\Theta_E = hv/k_B$ because $hv = k_B\Theta_E$ at the (fixed) Einstein temperature. Remember that $v$ is a parameter of the system in Einstein's theory, because the frequency of atomic vibrations is determined by the (essentially) Hooke's law forces tethering them to their individual lattice sites. Therefore $\Theta_E$ is not a new parameter; rather it is another way of writing the same parameter.

Any mutually agreed upon point on the individual heat capacity curves can be chosen as the normalizing temperature. Einstein chose $1325\,\mathrm{K}$ as

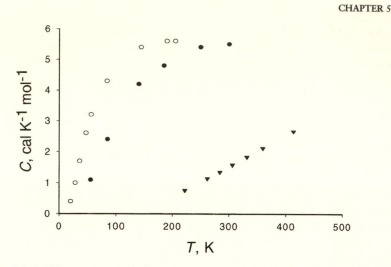

Figure 5.3.1a. Heat capacity of silver, (open circles), aluminum (solid circles), and diamond (triangles) versus $T$ in kelvins.

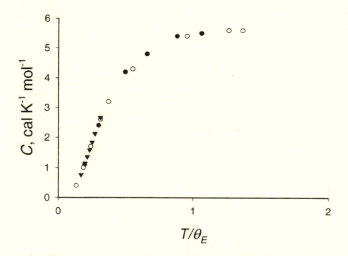

Figure 5.3.1b. The same data as in figure 5.3.1a, plotted as a function of the reduced variable $T/\Theta_E$. $\Theta_E$ is unique to each solid.

the normalizing temperature of diamond by arbitrarily taking the temperature at which $C$ is 1/4 of the way from the bottom of the $C_V$ versus $T$ curve (see problem 6.1). We have followed his method, arriving at 151 K and 284 K as the normalizing temperatures for silver and aluminum, respectively, in figure 5.3.1b.

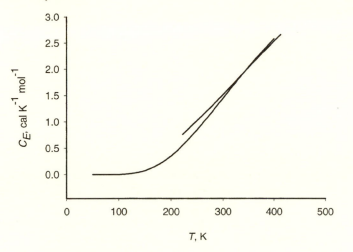

Figure 5.4.1. Discrepancies between the Einstein equation (lower curve) and the experimental curve for diamond at the low end of the temperature range. As experiments were carried out nearer absolute zero, these discrepancies became more pronounced.

## 5.4 Difficulties with the Einstein Theory

Comparison of the Einstein heat capacity curve with experimental data shows very good agreement in figures 5.2.1 and 5.3.1, but there are small discrepancies at temperatures well below $\Theta_E$ even for the original data set (diamond) (figure 5.4.1). Predictions from the Einstein model in this region are that the heat capacity is slightly smaller than the experimental values, that is, the predicted quanta are slightly larger than they are in reality.

If we had a model with essentially the same characteristics as the Einstein model, but one that allows a *distribution* of frequencies extending lower than the Einstein frequency, we would have a model that permits absorption of small heat quanta from the low-energy end of the quantal distribution, thus avoiding the Einstein problem at low temperatures. We would also have to place an *upper limit* on the number of modes to avoid the Rayleigh-Jeans predicament of infinite frequencies. Such a model should give better agreement with experiment than the Einstein model in the low-temperature region but, if the limiting frequency is appropriately chosen, it will still approach the limiting law of Dulong and Petit at high temperatures.

## PROBLEMS

**5.1.** Show that the Einstein heat capacity equation

$$C_E = \frac{3R \left( \dfrac{h\nu}{k_B T} \right)^2 e^{h\nu/k_B T}}{\left( e^{h\nu/k_B T} - 1 \right)^2}$$

reduces to the law of Dulong and Petit $C_E \to 3R \cong 25\,\mathrm{J\,K^{-1}\,mol^{-1}}$ at high temperatures $k_B T \gg h\nu$. Hint: use the series expansion $e^x = 1 + x + \cdots$ and ignore terms beyond $x$ in the sum for $x \ll 1$.

**5.2.** Einstein's predicted heat capacity falls below the experimental curve in figure 5.4.1. Give a qualitative argument showing that his predicted quanta are too large in this temperature region.

**5.3.** Consider a harmonic oscillator with only two energy levels, zero and $\varepsilon$. Call them state 0 and state 1. In general, there are many states and the probability that the oscillator will be in state $i$ is

$$p_i = C e^{-E/k_B T}$$

where $E$ is the energy of the $i$th level, and $C$ is a normalization constant required to make the total probability of finding the oscillator in one of the possible states $p = 1.00$. In this particular case, $E = $ either 0 or $\varepsilon$. Find $C$.

**5.4.** Having found the probability of state occupation, what is the average energy of the oscillator if it is observed very many times?

**5.5.** What are the limiting values of the average energy of the oscillator over very many observations as $T \to 0$ and as $T \to \infty$?

**5.6.** What is the molar heat capacity at constant volume, $C_V$, of the system of an Avogadro number of oscillators described in problem 5.3?

**5.7.** What is the Einstein heat capacity $C_E$ at the Einstein temperature $\Theta_E$?

**5.8.** Why did Einstein predict an absorption maximum for diamond at 11 microns (Einstein 1907)? (A micron is a micrometer, μm.)

**5.9.** Examine the function $(1/x)^2 e^{1/x} / (e^{1/x} - 1)^2$ between $x = 0$ and $x = 3$. Compare your result with equations 5.2.5 and 5.3.2.

# Six

## The Debye Equation

WHY SHOULD QUANTA have an upper limit on size? Why should they not increase to infinite size? How can one achieve a *distribution* of vibratory modes for a crystal? What would such a distribution look like? How would a distribution of quantal sizes improve agreement between experiment and theory for very low temperatures? What are the implications for the heat capacity of solids as a function of temperature?

### 6.1 The Debye Model

Picture a vibrating solid. For simplicity, let us take a cube of molar volume $V_m$. (This is a convenient assumption, not essential to the argument.) The normal modes of vibration must have nodes at each surface of the cube,

$$L = \frac{n\lambda}{2}, \qquad (6.1.1)$$

where $L$ is the length of the edge of the cube and $\lambda$ is the wavelength of allowed modes of vibration with wave numbers $n = 1, 2, 3, \ldots$. The factor 2 in the denominator appears because the fundamental mode of vibration ($n = 1$) has $\lambda = 2L$ (figure 6.1.1).

Following essentially the same logical argument as for the Rayleigh-Jeans equation (section 4.4), which gives $d\rho_N = (8\pi v^2/c^3)dv$ for a blackbody radiator, the *number density* $\rho_N$ of modes should increase with frequency as the square of the frequency times the frequency increment $dv$,

$$d\rho_N = av^2 dv, \qquad (6.1.2)$$

where $a$ is a constant of proportionality replacing $8\pi/c^3$ in the Rayleigh-Jeans equation. The total number density of modes is

$$\rho_N = \int av^2 dv \qquad (6.1.3)$$

with limits of integration that are yet to be specified.

The *upper limit* on the wavelength is twice the edge of a cubic crystal of molar volume (by hypothesis). This is a macroscopic dimension,

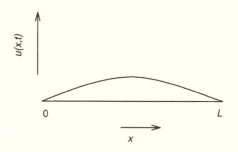

Figure 6.1.1. The fundamental mode of vibration in one dimension of a cubic solid.

essentially infinity on an atomic size scale; hence the *lower limit* on the *frequency* is the inverse of infinity, namely, zero.

The *lower limit* on *wavelength* is fixed because no vibration can occur with a wavelength that is smaller than the distance between atoms within the crystal. A wave in the space between atoms is meaningless because there is nothing there to vibrate.

●  ⌁⌁⌁  ●     Forbidden wave

A lower limit on $\lambda$ establishes an upper limit on $v$; hence the size of an allowed quantum of vibrational energy within the Debye crystal is limited by a fixed frequency called the *Debye cutoff frequency* $v_D$,

$$\varepsilon_D = h\, v_D. \qquad (6.1.4)$$

The integral 6.1.3 is over a spherical wave number space for large wave numbers comparable to figure 4.3.4. The radius of the sphere has a limit $v_D$. No matter what the distribution of frequencies may be, the number density of a molar volume must be the number of degrees of freedom of the atoms in a molar cube, $3N_A$,

$$\int_0^{v_D} av^2 dv = 3N_A. \qquad (6.1.5)$$

The model of standing waves in a crystal in which the atoms are not independent but transmit waves through their interatomic forces (i.e., the Debye model) is very like a model of sound waves propagated and reflected in a rectangular room or, on a smaller scale, within a rectangular block of metal. The *speed of sound* $v_s$ in a typical metal is a few thousand meters per second ($10^3$ m s$^{-1}$) and atomic separations are a hundred picometers or so, that is, $10^{-10}$ m, where 1 pm $= 10^{-12}$ m. This permits us to estimate a value of the Debye frequency by taking the minimum wavelength in a

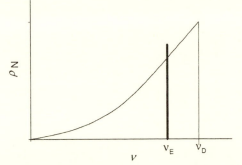

Figure 6.1.2. The Debye frequency spectrum. More complicated spectra are found in real crystals. The single, fixed Einstein frequency is shown as a solid line for comparison.

Debye solid to be of the same order of magnitude as the atomic separation. We get an estimate of $\nu_D \cong \nu_s/\lambda = 10^3/10^{-10} = 10^{13}$ Hz for the Debye frequency. An order of magnitude estimation of the size of the quantum of vibrational energy in a typical solid is

$$\varepsilon_D \cong h\nu_D = 10^{-34}10^{13} =\sim 10^{-21} \text{ J}.$$

Now that we have our distribution function $a\nu^2$, analogous to the Rayleigh-Jeans function, and the limits over which it is to be integrated, we can see that the Debye function is a parabola open upward over the range from 0 to $\nu_D$ as shown in figure 6.1.2. Different crystalline lattice types lead to functions more complicated than a simple parabola, but this is a technical matter that does not interrupt the logic of the presentation and need not concern us here.

We have now answered the first four questions at the beginning of this chapter. We know why there is a limit on quantal energy, we know about what the limit is, we have a distribution for $\rho_N(\nu)$, and we know what the distribution looks like. A distribution of number densities over frequencies is called a *spectrum* by analogy with the blackbody spectrum and the emission and absorption spectra studied in spectroscopy.

## 6.2 The Debye Equation

To answer the two remaining questions, we begin by performing the integration

$$\int_0^{\nu_D} a\nu^2 d\nu = \left.\frac{a\nu^3}{3}\right|_0^{\nu_D} = \frac{a\nu_D^3}{3}. \tag{6.2.1}$$

We know that

$$\int_0^{v_D} av^2 dv = 3N_A \tag{6.2.2}$$

hence

$$\frac{av_D^3}{3} = 3N_A \tag{6.2.3}$$

or

$$a = \frac{9N_A}{v_D^3}, \tag{6.2.4}$$

where $N_A$ is the Avogadro number for a molar volume. The molar number density of vibrational modes in a Debye solid is, from equation 6.1.2,

$$d\rho_N = \frac{9N_A}{v_D^3} v^2 dv. \tag{6.2.5}$$

Now, from Planck's equation, equation 4.5.23,

$$\bar{E} = \frac{E}{n} = \frac{hv}{\left(e^{hv/k_BT} - 1\right)} \tag{6.2.6}$$

and an infinitesimal in the energy density is

$$dU_D = Ed\rho_N = \bar{E}av^2 dv. \tag{6.2.7}$$

Combining these two equations, we arrive at

$$dU_D = \frac{9N_A}{v_D^3} \frac{hv}{\left(e^{hv/k_BT} - 1\right)} v^2 dv, \tag{6.2.8}$$

where the subscript $D$ denotes a physical property obtained from the Debye model in contrast to the Einstein model.

Rather than specifying a single frequency $v_E$ as in the Einstein model, we must now sum over all frequencies within a spectral range from 0 to $v_D$. Let us call each individual frequency within the spectral frequency range $v_j$. The contribution to the total heat capacity at each frequency *per atom per degree of freedom* is

$$C_j = \frac{k_B \left(\dfrac{hv_j}{k_BT}\right)^2 e^{hv_j/k_BT}}{\left(e^{hv_j/k_BT} - 1\right)^2}, \tag{6.2.9}$$

just as Einstein said it is, only now we have a multiplicity of contributions at different frequencies. Each $C_j$ contributes its bit to the heat capacity of the crystal. To find the total heat capacity of the crystal we sum over all contributions,

$$C_D = \sum_j C_j. \qquad (6.2.10)$$

This is difficult, but for a continuous solid the heat capacity is made up of very many, very small contributions $C_j$, times the weight of each $C_j$ in the distribution,

$$dC_D = C_j a v^2 dv \qquad (6.2.11)$$

so that the sum $C_D$ can be represented as an integral,

$$C_D = \int_0^{v_D} C_j a v^2 dv. \qquad (6.2.12)$$

Making the substitution $x = hv_j/k_B T$, Einstein's expression for $C_j$ becomes

$$C_j = \frac{k_B \left(\dfrac{hv_j}{k_B T}\right)^2 e^{hv_j/k_B T}}{\left(e^{hv_j/k_B T} - 1\right)^2} = k_B \frac{x^2 e^x}{(e^x - 1)^2}. \qquad (6.2.13)$$

The substitution $x = hv_j/k_B T$ also leads to

$$v^2 = \left(\frac{k_B T}{h}\right)^2 x^2$$

and

$$dv = \frac{k_B T}{h} dx.$$

Recalling that $a = 9N_A/v_D^3$, we can put the various pieces together to get

$$C_D = \frac{9N_A k_B}{v_D^3} \left(\frac{k_B T}{h}\right)^3 \int_0^{v_D} \frac{x^4 e^x}{(e^x - 1)^2} \, dx, \qquad (6.2.14)$$

where we bring $T$ out from under the integral sign because we are enquiring after the heat capacity of the crystal *at some fixed temperature*. But now notice that, although $hv = k_B T$ is true for any frequency, $hv_D = k_B T$ is true for only one specific frequency, the *Debye limiting frequency*. Let us call this unique temperature the *Debye temperature* and denote it $\Theta_D$, defined by $hv_D = k_B \Theta_D$. From this, we get $v_D = k_B \Theta_D/h$, and

$$v_D^3 = \left(\frac{k_B \Theta_D}{h}\right)^3.$$

Putting this into the Debye equation, the expression for $C_D$,

$$C_D = \frac{9N_A k_B}{v_D^3} \left(\frac{k_B T}{h}\right)^3 \int_0^{v_D} \frac{x^4 e^x}{(e^x - 1)^2} dx,$$

becomes

$$C_D = \frac{9N_A k_B}{\left(\frac{k_B \Theta_D}{h}\right)^3} \left(\frac{k_B T}{h}\right)^3 \int_0^{v_D} \frac{x^4 e^x}{(e^x - 1)^2} dx. \qquad (6.2.15)$$

The constants $k_B$ and $h$ cancel, and we remember that $N_A k_B = R$, leaving us with only

$$C_D = 9R \left(\frac{T}{\Theta_D}\right)^3 \int_0^{v_D} \frac{x^4 e^x}{(e^x - 1)^2} dx. \qquad (6.2.16)$$

To find the limits of integration we note that, as $v$ runs from 0 to $v_D$, $x$ runs over the range 0 to $h v_D / k_B T$, which is 0 to $\Theta_D / T$ because $\Theta_D = h v_D / k_B$. These, then, are the limits of integration leading to the final form of the Debye equation as

$$C_D = 9R \left(\frac{T}{\Theta_D}\right)^3 \int_0^{\Theta_D / T} \frac{x^4 e^x}{(e^x - 1)^2} dx. \qquad (6.2.17)$$

The integral is sometimes denoted simply as $D$ so a shorthand form of the Debye equation is

$$C_D = 9R \left(\frac{T}{\Theta_D}\right)^3 D. \qquad (6.2.18)$$

## 6.3 The Debye Temperature

Debye temperatures are not very different from Einstein temperatures and they follow a similar trend. For example, we used 151, 284, and 1325 K as the normalizing temperatures $\Theta_E$ for silver, aluminum, and diamond in Einstein's heat capacity plots. (Remember that these were chosen as arbitrary points on an Einstein plot; their absolute values are not important, what counts is the relation among them and with other values of $\Theta_E$ similarly chosen.) The Debye temperatures $\Theta_D = h v_D / k_B$ for Ag, Al, and $C_{\text{diamond}}$ are 215, 398, and 1840 respectively. These temperatures are (roughly) related to the Einstein temperatures we chose previously by $\Theta_D = 1.4 \Theta_E$.

Available experimental data, especially at low temperatures, were becoming more numerous and more accurate in the years following

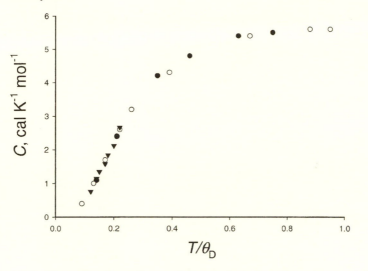

Figure 6.3.1. Normalization of experimental data for silver (open circles), aluminum (solid circles), and diamond (triangles) using the Debye temperature.

Debye's first 1912 paper, permitting determination of $\Theta_D$ for many solids. The reader may notice small discrepancies among Debye temperatures for the same solid in different tabulations because they were obtained from statistical fitting to different data sets (each with its own experimental error).

Another reason for discrepancies among Debye temperatures for the same solid is that there are different ways of obtaining $\Theta_D$. One is by fitting the Debye equation to an entire set of experimental results for heat capacities and another is by fitting a limiting form of the Debye equation to data only at the lowest temperatures. We shall use $\Theta_D$ obtained by the first method for the time being, and address the second method in more detail immediately below.

Once having a value for $\Theta_D$ and choosing $T/\Theta_D$ as our normalizing independent variable, all heat capacity curves coincide in figure 6.3.1, just as they did in the Einstein model when we plotted $C$ versus $T/\Theta_E$. In overall appearance, the Debye and Einstein curves are about the same. Examined in detail, both tend to the Dulong-Petit law at high temperatures, the Einstein curve fits experimental data slightly better than the Debye model at intermediate temperatures, and the Debye model fits experiment better than the Einstein curve at very low temperatures.

At the lowest temperatures, the Debye curve "tails off" toward zero slightly more than the abrupt Einstein curve. This is to be expected because it is only at the lowest temperatures that the lower end of the Debye spectrum (figure 6.1.2) influences heat capacity more than the upper end, where the Debye cutoff frequency functions essentially as the Einstein single frequency does.

## 6.4 The Integral $D$

The integral

$$D = \int_0^{\Theta/T} \frac{x^4 e^x}{(e^x - 1)^2} dx \tag{6.4.1}$$

in equation 6.2.17 is not known in analytical form but we can obtain its value by a numerical integration, for example, by Simpson's rule. Suppose we want to know the heat capacity of a metal at $\frac{1}{2}\Theta_D$. We integrate $D$ from a very small initial value to $\Theta_D/T = 2$. (We do not integrate from precisely 0 to $\Theta_D/T$ because the integrand is indeterminate at $0\,K$.) The result of a Simpson's rule integration from 0.0001 to 2 gives $D_2 = 2.201$. Thus,

$$C_D = 9RD\left(\frac{T}{\Theta_D}\right)^3 = 9R \times 2.201(\tfrac{1}{2})^3 = 20.59, \tag{6.4.2}$$

which, within the accuracy of the Debye model, is true for *all solids* at one-half the Debye temperature (law of corresponding states). This gives us another way of estimating an empirical value for $\Theta_D$. We need only find the temperature at which a solid has $C = 20.6$ by interpolation of experimental data, and double it. In the case of silver, the heat capacity is $20.6\,J\,K^{-1}\,mol^{-1}$ at about $115\,K$, which leads to $\Theta_D$ (silver) $= 230\,K$, in rough agreement with $\Theta_D$ (silver) $= 215\,K$ used in plotting figure 6.3.1.

## 6.5 Very-Low-Temperature Behavior of the Debye Equation

At very low temperatures, $\Theta_D/T$ can be regarded as $\infty$ because $T$ is very small. The integral $D$ becomes

$$D_\infty = \int_0^\infty \frac{x^4 e^x}{(e^x - 1)^2} dx \tag{6.5.1}$$

instead of

$$D = \int_0^{\Theta_D/T} \frac{x^4 e^x}{(e^x - 1)^2} dx. \tag{6.5.2}$$

In contrast to $D$, which has no analytical solution, the definite integral to an upper limit of infinity, $D_\infty$, is known:

$$D_\infty = \frac{4\pi^4}{15} = 25.976$$

(though it is not very easy to find in the handbooks). Alternatively, one can integrate numerically by Simpson's rule to a large upper limit, say 100. With this, equation 6.2.18 becomes

$$C_D = 9RD_\infty \left(\frac{T}{\Theta_D}\right)^3 = 9R\,(25.976)\left(\frac{T}{\Theta_D}\right)^3 = 1944 \left(\frac{T}{\Theta_D}\right)^3 \quad (6.5.3)$$

at very low temperatures. The unit of the constant 1944 must be $J\,K^{-1}\,mol^{-1}$ because both numerator and denominator in $(T/\Theta_D)$ have units of kelvins, so the ratio is unitless and $R$ is in units of $J\,K^{-1}\,mol^{-1}$.

As an example of low-temperature predictions of the Einstein and Debye functions, let us take, once again, 215 K as $\Theta_D$ for silver. For the integration to the upper limit $(T/\Theta_D) = 100, T = 100(215) = 215\,00$ K. Surely this is "infinite temperature" in any practical sense, so the integration to obtain equation 6.5.3 is valid. The Debye estimate of the heat capacity of Ag at, for example, 10 K is

$$C_D = 1944 \left(\frac{T}{\Theta_D}\right)^3 = 1944 \left(\frac{10}{215}\right)^3 = 0.196. \quad (6.5.4)$$

The experimental value is $0.199\,J\,K^{-1}\,mol^{-1}$. By contrast, the Einstein value at this temperature is about $5 \times 10^{-4}\,J\,K^{-1}\,mol^{-1}$ (nearly zero). We see the strength of the Debye model at low temperatures where frequencies at the low end of the parabolic number density spectrum in figure 6.1.2 dominate the heat capacity. The Einstein model has no lower end, indeed it has no spectrum at all. By any estimate, however, we are struck by how little heat, $q \cong C\Delta T$, silver absorbs for a small temperature rise in the vicinity of 10 K. Heat capacities of metals at low temperatures are very small but not zero.

Let us look more carefully at the low end of the heat capacity curve. Figure 6.5.1 shows that the Debye equation (upper curve) matches the experimental behavior of silver very well at temperatures below about 60 K where the Einstein curve (lower curve) fails. At temperatures that are low relative to most reaction temperatures, but not near absolute zero, the Einstein curve approaches the experimental data better than it does at the very lowest temperatures. By the time we have reached 100 K the Debye curve overestimates the heat capacity of silver, while the Einstein curve comes closer to the experimental points, for example, at $T \sim 85$ K in figure 6.5.1. At temperatures higher than 100 K the Einstein curve will

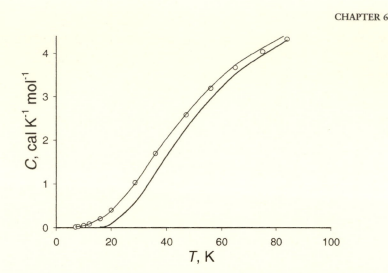

Figure 6.5.1. The Debye approach to absolute zero for silver. The Einstein equation is given by the lower curve. Open circles are experimental points.

be a better representation of the experimental data for silver. The curves cross just above 100 K and cross yet again at a higher temperature. Above 100 K, discrepancies between the two theoretical curves and experimental data are negligible.

## 6.6 The Speed of Sound in Solids

At first thought, there might not seem to be much connection between the speed of sound in a metal and its heat capacity. Sound is a wave displacement of alternating compressions and decompressions in whatever material the sound travels through. When a "ping" is struck at one end of a metal bar, a longitudinal compression pulse travels through the bar and is reflected at the other end. Just as in a vibrating guitar string, standing waves are or are not maintained according to whether the boundary conditions of length, mass and restoring force are or are not met.

.   .   .   .   .   ...   .   .   .   .   .   ...   .   .   .   .   ...   .   .   .   .

→

The Debye model of heat capacity is based on the resistance to displacement of atoms within a solid restrained by the aggregate of all neighboring forces holding them to lattice points within the atomic lattice. Larger collaborative groups of atoms (larger $m$) vibrate at lower frequencies: $v = (1/2\pi)\sqrt{k/m}$, as in section 2.1, where $v = 2\pi\omega$. Different modes of atomic motion lead to different frequencies in solids just as different modes of motion lead to different sound frequencies in air. Debye reasoned that collaborative motion among atoms in a block of solid metal brings about a disturbance involving *groups* of atoms with a fundamental and overtones that are analogous to sound waves progressing through a rectangular chamber or room. As such, the number density of vibratory modes is given by the Rayleigh-Jeans equation, equation 4.3.9 (which was developed as part of Rayleigh's studies on acoustics long before quantum theory)

$$d\rho_N = \frac{4\pi v^2}{v_s^3} dv \tag{6.6.1}$$

where $v_s$ is the velocity (speed) of sound in a chamber or a block of metal.

The number density is $d\rho_N = dN/V$ where $V$ is the volume of the chamber or block. This leads to

$$dN = \frac{4\pi v^2 dv}{v_s^3} V \tag{6.6.2}$$

for the increment in modes of vibration $dN$ with an increment in frequency $dv$ and

$$\int dN = \int \frac{4\pi V}{v_s^3} v^2 \, dv \tag{6.6.3}$$

for the total number of allowed modes. This total $\int dN$ is $3N$, the number of degrees of freedom of all the atoms in the block by the same reasoning that led to equation 6.1.5. The upper limit of integration on the right is $v_D$ by the reasoning that preceded equation 6.1.4. If we arbitrarily choose a *molar* volume of metal $V_m$, the integral is $3N_A$ where $N = N_A$, the Avogadro number. This leads to

$$3N_A = \frac{4\pi}{v_s^3} V_m \int_0^{v_D} v^2 \, dv = \frac{4\pi}{3v_s^3} V_m v_D^3. \tag{6.6.4}$$

Now

$$v_D^3 = 3N_A \frac{3v_s^3}{4\pi V_m} = \frac{9}{4\pi} \left(\frac{N_A}{V_m}\right) v_s^3 \tag{6.6.5}$$

or

$$v_D = \left[\frac{9}{4\pi} \left(\frac{N_A}{V_m}\right)\right]^{1/3} v_s. \tag{6.6.6}$$

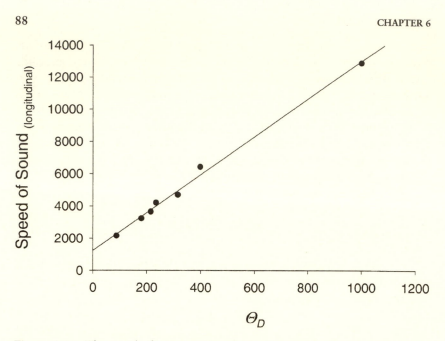

Figure 6.6.1. The speed of propagation of sound versus the Debye temperature obtained from heat-capacity curves for metals. The metals are Pb, Au, Ag, Zn, Cu, Al, and Be.

Thus, one can calculate $\Theta_D = h\, v_D/k_B$ from the speed of sound in elastic solids,

$$\Theta_D = \frac{h}{k_B}\left[\frac{9}{4\pi}\left(\frac{N_A}{V_m}\right)\right]^{1/3} v_s. \qquad (6.6.7)$$

This leads us to believe that $\Theta_D$ should increase in a linear way with $v_s$, and we see in figure 6.6.1 that it does.

For accurate calculations, we encounter a complication that has been ignored so far. Sound travels in a solid with three velocities, one *longitudinal*, in the direction of the sound wave, and two *transverse* velocities corresponding to the two dimensions perpendicular to propagation of the wave.

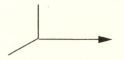

If these speeds were the same, we could lump them together as one $v_s$ but in general, the two transverse speeds differ from the longitudinal speed.

This means that the Rayleigh "sphere" containing all the wave numbers less than some selected frequency $v_D$ is not really a sphere at all in Debye theory but, in the case of equal transverse speeds, it is an ellipsoid of revolution. One way of compensating for the difference in longitudinal and transverse speeds is to substitute the cube root of the harmonic mean of the cubed speeds for $v_s$ in equation 6.6.7

$$\Theta_D = \frac{h}{k_B} \left[ \frac{9}{4\pi} \left( \frac{N_A}{V_m} \right) \right]^{1/3} \left( \frac{1}{\frac{1}{v_{s,l}^3} + \frac{2}{v_{s,t}^3}} \right)^{1/3}, \qquad (6.6.8)$$

This complication changes the shape of the spectral curve in figure 6.1.2. The peak is not as sharp as the Debye approximation would have it, rather it is a superposition of similar curves with $v_{s,l} \neq v_{s,t}$, where the subscripts $s, l$ and $s, t$ designate the longitudinal and transverse speeds. Different forms of the crystal unit cell (face centered cubic, body centered cubic, etc.) also alter the Debye spectrum from the simple form in figure 6.1.2. All of these complications serve to remind us that the Debye spectrum in figure 6.1.2 is an approximation. Nonetheless, considering its simplicity, the Debye prediction of a linear relation between $v_s$ and $\Theta_D$ fits experimental observations of $v_s$ rather well. For example, choosing the velocity of the longitudinal wave for our independent variable, figure 6.6.1 shows $\Theta_D$ calculated from heat capacity curves versus $v_{s,l}$.

Finally, if the heat capacity of a mole of metal is

$$C_D = R \sum_j \left( \frac{\Theta_D}{T} \right)^2 \frac{e^{\Theta_D/T}}{\left( e^{\Theta_D/T} - 1 \right)^2}, \qquad (6.6.9)$$

where each of many $j$ terms corresponds to a harmonic vibration in the solid sample, we can replace the sum with an integral,

$$C_D = \int_0^{\Theta_D} \frac{k_B x^2 e^x}{(e^x - 1)^2} 4\pi v^2 V_m \left( \frac{1}{v_{s,l}^3} + \frac{2}{v_{s,t}^3} \right)^{1/3} dv \qquad (6.6.10)$$

or

$$C_D = \frac{9R}{x_0^3} \int_0^{x_0} \frac{x^4 e^x}{(e^x - 1)^2} dx, \qquad (6.6.11)$$

where $x = hv/k_B T$ and $x_0 = h v_D/k_B T$. Equation 6.6.11 is just equation 6.2.14 or 6.2.16 modified so that the speed of sound enters into it through $x_0$ and $v_D$. This makes the connection between heat capacity and the speed of sound in a solid metal.

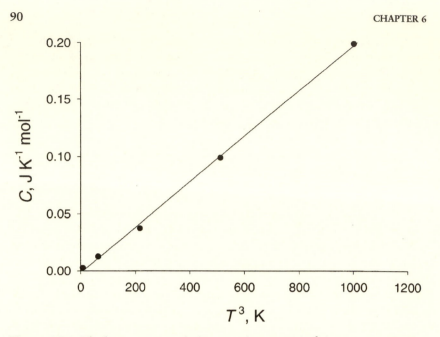

Figure 6.7.1. The heat capacity of silver as a function of $T^3$ from 1 to 10 K. The slope is $1.99 \times 10^{-4}$ J K$^{-4}$ mol$^{-1}$.

## 6.7 The Debye Third-Power Law

If $C_D = 1944 \, (T/\Theta_D)^3$, then $C_D$ *versus* $T^3$ should be a straight line with a slope of $1944/\Theta_D^3$, but only at very low temperatures. Figure 6.7.1 has a slope of $1.99 \times 10^{-4}$ J K$^{-4}$ mol$^{-1}$. Thus, for silver,

$$\Theta_D = \left( \frac{1944}{1.99 \times 10^{-4}} \right)^{1/3} = 214 \, \text{K} \tag{6.7.1}$$

which is a good approximation to the previous values of 215 K (section 6.3) and 230 K (section 6.4).

## 6.8 Third-Law Entropies

Knowing that the heat capacity of solids goes up as $T^3$,

$$C_P = AT^3, \tag{6.8.1}$$

one can find $A$ from a known value of $C_P$ at, say, 10.0 K,

$$A = \frac{C_P}{1000} \quad \text{at } 10.0 \, \text{K}, \tag{6.8.2}$$

where we are now specifying the heat capacity *at constant pressure*, $C_P$. In the case of liquids and especially vapors, the heat capacity at constant pressure differs from the heat capacity at constant volume, $C_V$, by a significant amount, and we must make the distinction that we have not made up to now. With the value of $A$, one can calculate $C_P$ of the solid over the range from $0\,K$ to $10.0\,K$, a temperature region that is experimentally accessible only with difficulty. When this is done, the *absolute entropy* is found at 10 K from the integral

$$S = \int_0^{10\,K} \frac{C_P}{T} dT = \int_0^{10\,K} \frac{AT^3}{T} dT = \int_0^{10\,K} AT^2 dT$$

$$= A \frac{T^3}{3} \bigg|_0^{10\,K} = \frac{A}{3} (1000). \tag{6.8.3}$$

Further experimental measurements yield the absolute entropy at any temperature. Over the temperature range in which the substance is a solid, numerical integration,

$$\Delta S = \int_{10K}^{T_2} \frac{C_P}{T} dT, \tag{6.8.4}$$

gives the entropy change over the interval $10.0\,K$ to $T_2$ which when added to the Debye third-law entropy is the absolute entropy at $T_2$. If the solid undergoes a phase change (e.g., a change in crystalline form), a discontinuity in the curve of $C_P/T$ versus $T$ is observed. The same is true of changes in state from solid to liquid and liquid to vapor. The entropy increase for each phase change taking place between $0\,K$ and a selected temperature $T$ is added to the integrals of $C_P/T$ over the temperature intervals $T_i$ to $T_j$ between phase changes to find the absolute entropy at the selected temperature,

$$S = \sum \int_{T_i}^{T_j} \frac{C_P}{T} dT + \sum \frac{\Delta_{\text{phase}} H}{T_{\text{phase}}}, \tag{6.8.5}$$

say, $298.15\,K$. With the absolute entropies of all components of a chemical reaction, one can obtain the entropy of reaction at any selected temperature of reaction:

$$\Delta_r S = \sum S(\text{prod}) - \sum S(\text{react}). \tag{6.8.6}$$

Entropies of reaction, in combination with enthalpies of reaction $\Delta_r H$, yield information of immense practical importance through the Gibbs formula for the chemical potential of reaction, $\Delta_r G = \Delta_r H - T\Delta_r S$, and the relation between the *standard free energy change* $\Delta_r G^0$ and the equilibrium constant of a chemical or physical reaction,

$$\Delta_r G^0 = -RT \ln K_{eq}. \qquad (6.8.7)$$

## PROBLEMS

**6.1.** In what may be the most widely reproduced data set in all of physical science (figure 5.2.1), H. F. Weber, in 1875, found $C_V = 6.24 \, \mathrm{J \, K^{-1} \, mol^{-1}}$ (converted to modern units) at $T = 331.3 \, \mathrm{K}$ for diamond. This point is just one-quarter of the way up the $C_V$ versus $T$ curve. Einstein arbitrarily chose it as his normalizing temperature for diamond. Find $\Theta_E$. Using the empirical generalization $\Theta_D = 1.4 \Theta_E$, estimate $\Theta_D$ and compare it with the accepted value of $\Theta_D = 1860 \, \mathrm{K}$.

**6.2.** Using a commercial plotting program (for example, Mathcad©), plot the Einstein function from $T = 0$ to $1600 \, \mathrm{K}$ for diamond.

**6.3.** Do the same for the Debye function. Now plot both curves on the same graph.

**6.4.** Magnify the region at the lower end of the two curves from problem 6.9.3 to show the area bounded by heat capacities from 0 to $10 \, \mathrm{K}$ and temperatures from 0 to $400 \, \mathrm{K}$.

**6.5.** (a) Using Simpson's rule, write a program in BASIC that enables you to generate a 16-entry table of the definite integrals $D$ (section 6.4) from approximately 0 to upper limits $\Theta_D/T$ of 1 to 16. Alternatively, you may want to use one of the many commercial programs that evaluate definite integrals over different limits.

  (b) Take $\Theta/T$ to a large value, say 100, to see if you can verify $D = 4\pi^4/15 = 25.976$. Be aware of limited precision (bit numbers) and rounding error.

**6.6.** The density $d$ of copper is $8.92 \, \mathrm{g \, cm^{-1}}$. What is its molar volume $V_m$?

**6.7.** Given the result from problem 6.6, what is the average volume occupied by a Cu atom in copper metal? What is the minimum interatomic distance?

**6.8.** Having $V_m/N_A$ from problem 6.7, find the Debye frequency $\nu_D$ and Debye temperature $\Theta_D$ from the weighted mean speed of sound in metallic copper which has $\nu_{s,t} = 2100 \, \mathrm{m \, s^{-1}}$ and $\nu_{s,l} = 4700 \, \mathrm{m \, s^{-1}}$. Don't forget that there are two transverse components and only one longitudinal component.

**6.9.** As we shall soon see, helium has many unique physical properties, especially at low temperatures. Solid helium has an interatomic separation of about 300 pm and a Debye temperature of about $\Theta_D =$ 30 K. Estimate the speed of sound in solid He and compare it with the speed of sound in air (about $325\,\mathrm{m\,s^{-1}}$).

6.10. The heat capacity of gold at 10.0 K is $0.431\,\mathrm{J\,K^{-1}\,mol^{-1}}$. Calculate the entropy of gold at 10.00 K.

# Seven

## Quantum Statistics

In 1905, THE PHOTOELECTRIC EFFECT was explained by Einstein (Nobel prize, 1922), who pictured a metal as an atomic lattice immersed in a sea of electrons which are rather loosely attracted to the lattice by its overall positive charge. Rejecting the wave theory of light for this phenomenon at least, he proposed that, in photoelectric conduction, light *particles* of energy $E = hv$ strike the surface of the electron sea and "splash" out electrons, but only if they have enough energy to overcome the attraction of the positive atomic lattice for the negative electrons. The energy of light particles is known, he said, from Planck's theory of blackbody radiation. These light particles were later called *photons*.

## 7.1 The Photoelectric Effect

When light falls on the surface of many metals, for example, the alkali metals, electrons may be released from the metal. If so, the circuit in figure 7.1.1 is completed and current flow is detected by the galvanometer. This is called the *photoelectric effect*.

Current is detected only when electrons flow across the cathode-anode gap in figure 7.1.1. According to wave theory prevalent at the time, increasing the *intensity* of light should gradually increase the amplitude of oscillation of electrons within the cathode, eventually causing them to oscillate so violently that they break away from their nuclei and leave, causing current to flow. Experiment contradicted this picture. Among other contradictions, when light impinging on the metal cathode is below a threshold *frequency*, current does not flow even if the light is made more intense. Instead, increasing the frequency of incident light brings about current flow.

If the polarity of the cell in figure 7.1.1 is reversed, the phototube still works but it requires light of a higher frequency to bring about current flow. Now the ejected electron is opposed by the electric field. The potential energy of this field is the charge on the electron times the potential difference between the electrodes. Even well above the threshold frequency, a reverse potential can be found that is sufficient to stop electrons thus preventing current flow. This potential is called the *stopping potential $V_s$*.

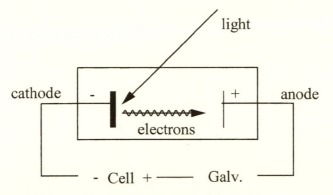

Figure 7.1.1. Schematic drawing of a phototube. When light of frequency $v$, strikes the cathode surface, electrons are emitted and the circuit is completed as detected by the galvanometer Galv. For this to happen $v$ must be above a *threshold frequency* characteristic of the cathode metal.

One can measure the maximum kinetic energy of the ejected electrons by adjusting the stopping potential until it just prevents current flow,

$$-eV_s = \tfrac{1}{2}mv^2 = E_{kin}, \qquad (7.1.1)$$

where the negative sign arises because the charge on the electron is negative.

Mathematically, the Einstein model of monochromatic light particles striking the surface of an electron sea, splashing out electrons or failing to do so, according to whether the photon energy is more or less than the threshold energy, is consistent with a function $E_{kin}$ versus $hv$ having a horizontal intercept equal to the threshold energy in units of $hv$. Beyond the threshold, excess energy over and above the energy necessary to drive electrons away from the positive ion lattice is communicated to the electrons as energy of motion $E_{kin} = \tfrac{1}{2}mv^2$. If the model is accurate, this energy is communicated by photons having energy $E = hv$. The function $E_{kin}$, resulting from a knowledge of $-eV_s$, versus $v$ should be linear, and it should have a slope of $h$. In 1905, however, the charge on the electron was not known, so the proportionality constant between $-eV_s$ and $v$ could not be found.

R. S. Millikan, at the culmination of a long and difficult series of experiments by himself and others, determined the charge on the electron and found, in 1916, that the slope of the linear function in figure 7.1.2 is indeed Planck's constant, $6.57 \times 10^{-27}$ erg s, within the combined experimental uncertainty of the blackbody radiation spectrum and his method. The presently accepted value of $h$ is $6.626 \times 10^{-27}$ erg s $= 6.626 \times 10^{-34}$ J s.

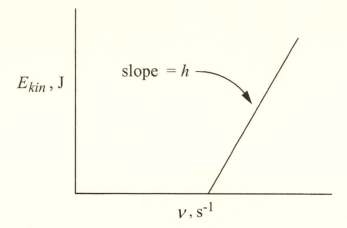

Figure 7.1.2. Determination of Planck's constant. The maximum kinetic energy of emitted electrons is found from the minimum stopping potential. The modern unit of Planck's constant $h$ is J s.

Notice the units: J per $s^{-1}$ = J s, in other words, the reciprocal of reciprocal seconds is seconds. Planck's constant is now recognized as one of the fundamental constants of the physical universe.

Initially, Einstein's use of "light particles" encountered quite a bit of opposition because physicists were well aware of the many classical experiments (diffraction, for example) proving that light has *wavelike* properties. It was some time before the mainstream of physicists got used to the idea that photons have *both* wave and particle properties.

## 7.2 The Photon Gas

Within the context of a photon theory of light, let us return to the thermostated "evacuated chamber" of chapter 3, which we claimed is not empty but which contains *radiation* at any temperature above absolute zero on the kelvin scale. Light, like thermal radiation, is electromagnetic radiation, which can be regarded either as an oscillation in the electromagnetic field or as moving light particles (photons). Photons transmit electromagnetic energy as Einstein showed by his work on the photoelectric effect. From this point of view, an evacuated chamber at thermal equilibrium above 0 K does not contain nothing, it contains *photons* (figure 7.2.1). The existence of a field implies the existence of particles.

After thermal equilibrium has been established within an insulated box at some temperature $T \neq 0$ K, constant temperature within the box

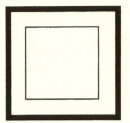

Figure 7.2.1. An evacuated, thermostated box contains photons at $T > 0$.

implies that the walls of the box must be emitting and absorbing photons at the same rate. The equilibrium number of photons within the box is determined only by the temperature, which implies that in order to change the number of photons within the box, one does not need to insert or withdraw photons; they can be *created or destroyed* by changing the temperature of the box.

If photons are particles, they should obey at least some of the rules of particles. It should be possible to treat a perfectly evacuated container at $T \neq 0$ K filled with a photon gas as though it were a chamber filled with an ordinary monatomic gas. The rules for a photon gas are not exactly the same as the rules for an ideal gas, however, because photons, unlike gas particles, are not conserved.

## 7.3 Bose's Letter to Einstein

In the summer of 1924, Einstein received a curious letter from S. N. Bose of Dacca University, India. In it was a manuscript in English, previously rejected by the *Philosophical Magazine*, with a request for help in translating it into German, a language in which Bose was not adept. With this was also a request that Einstein forward the translated manuscript to the *Zeitschrift für Physik* with his recommendation for publication.

This second request was not as unusual as it sounds. It was the custom for some journals to accept manuscripts from junior scientists only after the manuscript had been sent to them by senior scientists with a positive recommendation. This is not very different from our present-day custom of submitting manuscripts to regional editors who send them out for peer review and then on to the editor-in-chief with a recommendation to publish or not to publish.

Einstein immediately recognized the importance of Bose's short paper (four journal pages). He translated it and sent it to *Zeitschrift für Physik*, where it was published in late 1924. In it Bose criticized earlier derivations of the Planck equation

$$\rho_v \, dv = (8\pi v^2 \, dv/c^3)E \tag{7.3.1}$$

on the ground that it is derived from inconsistent (or even contradictory) origins because the term in parentheses comes from classical theory and $E$ (strictly $\bar{E}$) comes from quantum theory. Derivation from a model of a *statistical distribution* of different "sorts of quanta" over a manifold of quantum levels separated from one another by equal increments in energy avoided this dichotomy, said Bose. In effect, Bose was replacing Planck's or Einstein's semiclassical model of many harmonic oscillators interacting with radiation with a model consisting of *one* harmonic oscillator having infinitely many equally spaced quantum levels, over which identical photons are distributed in a way that is calculable by statistical methods provided a quarter-century earlier by J. W. Gibbs. Indeed, one can dispense with the harmonic oscillator entirely and picture the *electromagnetic field* within the blackbody chamber as being quantized with equally spaced energy levels. This connection between a quantized field and a particle (photon) was to have a profound influence on physics for the rest of the century. Although Gibbs's statistical methods were arrived at with classical systems in mind, they are sufficiently general that neither Gibbs's nor Bose's derivations depends on a classical model. Einstein saw the generality of Bose's method and soon developed it into the subdivision of *quantum statistics* that we now call *Bose-Einstein statistics*.

## 7.4 The Quantum Harmonic Oscillator

We have seen the classical Boltzmann distribution for energy as a continuous function (equation 2.11.2)

$$\frac{dn_i}{dn_0} = e^{-\varepsilon_i/k_BT}, \tag{7.4.1}$$

where $\varepsilon_i$ is the energy relative to a defined energy level $\varepsilon_0 = 0$. The quantum derivation rests solidly on the concept of a *discontinuous* energy distribution in which each energy level is separated from its neighbors by an amount of energy $\Delta\varepsilon = hv$ (figure 7.4.1); nevertheless, it arrives at a very similar distribution.

The ratio of infinitesimals $dn_i/dn_0$ in the continuous Boltzmann expression is replaced by finite numbers $n_i$ and $n_0$,

$$\frac{n_i}{n_0} = e^{-\varepsilon_i/k_BT} = e^{-\beta\varepsilon_i} \tag{7.4.2}$$

*etc.*

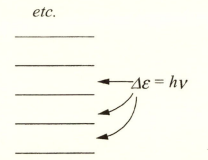

Figure 7.4.1. Quantized energy levels. Equally spaced energy levels like these are sometimes called an energy ladder.

where, for notational simplicity, we use the common abbreviation $\beta = 1/k_B T$. The number of particles at any level is

$$n_i = n_0 e^{-\beta \varepsilon_i}. \qquad (7.4.3)$$

A particle at level $\varepsilon_2$ contributes twice as much energy as a particle at $\varepsilon_1$, two particles at $\varepsilon_2$ contribute four times as much energy as one at $\varepsilon_1$, and so on. The total number of particles summed over nondegenerate levels is

$$N = \underset{\text{levels}}{\sum} n_i = \underset{\text{levels}}{\sum} n_0 e^{-\beta \varepsilon_i} = n_0 \underset{\text{levels}}{\sum} e^{-\beta \varepsilon_i} \qquad (7.4.4)$$

and the total energy is

$$E = \underset{\text{levels}}{\sum} n_i \varepsilon_i = \underset{\text{levels}}{\sum} \varepsilon_i n_0 e^{-\beta \varepsilon_i} = n_0 \underset{\text{levels}}{\sum} \varepsilon_i e^{-\beta \varepsilon_i}. \qquad (7.4.5)$$

For a Boltzmann distribution of particles over nondegenerate energy levels, the probability $p_i$ of finding a particle at the $i$th level is

$$p_i = \frac{n_i}{N} = \frac{n_0 e^{-\beta \varepsilon_i}}{n_0 \underset{\text{levels}}{\sum} e^{-\beta \varepsilon_i}} = \frac{e^{-\beta \varepsilon_i}}{\underset{\text{levels}}{\sum} e^{-\beta \varepsilon_i}} = \frac{e^{-\beta \varepsilon_i}}{q}, \qquad (7.4.6)$$

where we call the sum over levels in the denominator $q$ the *particle partition function*

$$q = \underset{\text{levels}}{\sum} e^{-\beta \varepsilon_i}. \qquad (7.4.7)$$

Our vibrational model consists of a one-dimensional energy space with a manifold of distinct quantum levels at equal intervals. The Boltzmann distribution function for $N$ particles over the space at constant temperature gives

$$n_i = N\frac{e^{-\beta\varepsilon_i}}{q}. \tag{7.4.8}$$

The particle partition function for evenly spaced energy levels is

$$q = \sum_{\text{levels}} e^{-\beta\varepsilon_i} = 1 + e^{-\beta\varepsilon} + e^{-2\beta\varepsilon} + e^{-3\beta\varepsilon} + \cdots, \tag{7.4.9}$$

whence

$$q = \frac{1}{1 - e^{-\beta\varepsilon}} \tag{7.4.10}$$

by manipulation of the sum as previously shown (equation 4.5.11). The subscript on $\varepsilon$ is not needed in the expression for $q$ because the levels are equidistant, that is, $\varepsilon_1 = \varepsilon$, $\varepsilon_2 = 2\varepsilon$, and so on. Taking natural logarithms,

$$\ln q = -\ln\left(1 - e^{-\beta\varepsilon}\right). \tag{7.4.11}$$

The vibrational energy is

$$E = \sum_{i=0}^{\infty} n_i\varepsilon_i \tag{7.4.12}$$

for the harmonic oscillator. The number of particles distributed over the manifold is

$$N = \sum_{i=0}^{\infty} n_i = \sum_{i=0}^{\infty} n_0 e^{-\beta\varepsilon_i} = n_0 \sum_{i=0}^{\infty} e^{-\beta\varepsilon_i} = n_0 q \tag{7.4.13}$$

because $n_i/n_0 = e^{-\beta\varepsilon_i}$. We can see that $q$ tells us something about how the particles are distributed over the energy-level manifold. If $q$ is very small (with a lower limit of 1), $N = n_0 q$ shows us that $N \cong n_0$, that is, most of the particles are in the lowest level. If $q$ is large, $n_0 \ll N$ and particles are not concentrated at the $n_0$ level. They may be concentrated at some other level $n_i \neq n_0$ or they may be widely distributed over the energy-level manifold. Because $\beta = 1/k_B T$, we expect $q$ to be small at low temperatures where the negative exponent in $q = \sum_{\text{levels}} e^{-\beta\varepsilon_i}$ becomes large. We shall soon see that, for Fermi-Dirac and Bose-Einstein distributions, the low-temperature condition is important.

## 7.5 The Total Vibrational Energy

The total vibrational energy of all the $N$ particles is

$$E = \sum_{i=0}^{\infty} n_i \varepsilon_i = \sum_{i=0}^{\infty} N \frac{e^{-\beta \epsilon_i}}{q} \varepsilon_i = \frac{N}{q} \sum_{i=0}^{\infty} \varepsilon_i e^{-\beta \epsilon_i} \qquad (7.5.1)$$

but

$$-\frac{dq}{d\beta} = -\frac{d}{d\beta} \sum e^{-\beta \varepsilon_i} = \sum \varepsilon_i e^{-\beta \varepsilon_i};$$

hence

$$E = \frac{N}{q} \sum_{i=0}^{\infty} \varepsilon_i e^{-\beta \varepsilon_i} = \frac{-N}{q}\left(\frac{dq}{d\beta}\right) = -N\left(\frac{\dfrac{dq}{q}}{d\beta}\right) = -N\left(\frac{d \ln q}{d\beta}\right).$$

$$(7.5.2)$$

This gives

$$E = -N\left(\frac{d \ln q}{d\beta}\right) = N\left(\frac{d \ln (1 - e^{-\beta \varepsilon})}{d\beta}\right) = N\varepsilon \frac{e^{-\beta \varepsilon}}{1 - e^{-\beta \varepsilon}} \qquad (7.5.3)$$

where we recall that $\ln q = -\ln\left(1 - e^{-\beta \varepsilon}\right)$ and

$$\frac{d}{d\beta} \ln (1 - e^{-\beta \varepsilon}) = \frac{1}{(1 - e^{-\beta \varepsilon})} \frac{d(1 - e^{-\beta \varepsilon})}{d\beta} = \frac{\varepsilon e^{-\beta \varepsilon}}{1 - e^{-\beta \varepsilon}}.$$

Some convenience substitutions yield

$$E = \frac{Nk_B hv}{k_B} \frac{e^{-\beta \epsilon}}{1 - e^{-\beta \epsilon}} = \frac{Rhv}{k_B} \frac{e^{-\beta \varepsilon}}{1 - e^{-\beta \varepsilon}} \qquad (7.5.4)$$

where $\varepsilon = hv$, or, setting $u = hv/k_B T = \beta \varepsilon$,

$$E = RT \frac{hv}{k_B T} \frac{e^{-\beta \varepsilon}}{1 - e^{-\beta \varepsilon}} = RTu \frac{1}{e^u - 1} \qquad (7.5.5)$$

per degree of freedom, for a molar quantity.

## 7.6 Heat Capacity

Given that

$$E = RTu(e^u - 1)^{-1} \qquad (7.6.1)$$

and that the heat capacity is the derivative of the energy with respect to temperature,

$$C = \frac{dE}{dT} = \frac{d}{dT} RTu(e^u - 1)^{-1},$$

we get three terms

$$C = Ru(e^u - 1)^{-1} + RT(e^u - 1)^{-1} \frac{du}{dT} + RTu \frac{d(e^u - 1)^{-1}}{dT}. \quad (7.6.2)$$

Note that because $u = hv/k_BT$, $du/dT = -hv/k_BT^2 = -u/T$, and

$$\frac{d(e^u - 1)^{-1}}{dT} = -\frac{e^u}{(e^u - 1)^2} \frac{du}{dt},$$

$$C = Ru(e^u - 1)^{-1} + RT(e^u - 1)^{-1} \left(\frac{-u}{T}\right) - RTu \frac{e^u}{(e^u - 1)^2} \frac{-u}{T}.$$

$$(7.6.3)$$

The first two terms cancel, leaving

$$C_{vib} = Ru^2 \frac{e^u}{(e^u - 1)^2} \quad (7.6.4)$$

for one mole, per degree of freedom. We write the heat capacity as $C_{vib}$ as a reminder that only the vibrational energy and the vibrational contribution to the heat capacity have been considered. The Einstein equation can be written

$$C_{vib} = 3R (\beta hv)^2 \frac{e^{\beta hv}}{\left(e^{\beta hv} - 1\right)^2} \quad (7.6.5)$$

for an isotropic solid with three degrees of freedom. This is the same equation that Einstein obtained by different means in 1907 (section 5.2),

$$C_E = \frac{3R \left(\frac{hv}{k_BT}\right)^2 e^{hv/k_BT}}{\left(e^{hv/k_BT} - 1\right)^2}.$$

Use of a classical Boltzmann statistical distribution over a discontinuous energy-level manifold permits the assumption that particles do not influence each other (except for free exchange of energy, which is called "weak interaction"). In the independent particle approximation, there is no limit on the number of particles that can exist in a state.

## 7.7 Bosons and Fermions

Nature has been astonishingly economical with respect to the kinds of subatomic particles. So far as we know, all particles are either *bosons* or *fermions*. Fermions, named after the Italian physicist Enrico Fermi, are familiar to chemists because the electron, which accounts for almost all of chemistry, is a fermion.

Electrons cannot have all four quantum numbers the same, which is a way of saying that they do not ever occupy the same state. Bosons can exist in the same quantum state. When the most probable distributions of fermions and bosons are worked out for the harmonic oscillator model, they give results that are different from the classical model for distinguishable objects because they are *indistinguishable*. Bosons and fermions give results that are different from each other as well, because of the different ways they occupy available states.

We shall find three distributions governing the way particles occupy a spectrum of energy levels like the ladder in figure 7.4.1. In the more general case, levels need not be equidistant and levels may be *degenerate*, that is, two or more states may have the same energy.

CASE 1. THE CLASSICAL BOLTZMANN DISTRIBUTION

$$n_i = \frac{1}{e^{-\alpha}e^{\beta\varepsilon}} = e^{\alpha}e^{-\beta\varepsilon} \tag{7.7.1}$$

which is a limiting form of

CASE 2. THE FERMI-DIRAC DISTRIBUTION (OF FERMIONS)

$$n_i = \frac{1}{e^{-\alpha}e^{\beta\varepsilon} + 1} \tag{7.7.2}$$

or

CASE 3. THE BOSE-EINSTEIN DISTRIBUTION (OF BOSONS)

$$n_i = \frac{1}{e^{-\alpha}e^{\beta\varepsilon} - 1}. \tag{7.7.3}$$

Thinking qualitatively about these distributions, one expects to find that the differences are important only at very low temperatures. Restrictions on multiple occupancy of orbitals do not have much effect at high temperatures because there is enough thermal energy available to particles that they can occupy a wide distribution of states, well scattered over the energy-level spectrum. Particles do not double up because there is no need to. It is only at low temperatures that particles tend to crowd into the lowest available orbitals. Bosons can double up, triple up, and so on, but fermions cannot. In the limit of 0 K, bosons can crowd into the lowest available level but fermions must occupy a number of states equal to the number of fermions, one to each. This leads to different mathematical distributions and a difference in low-temperature physical behavior.

Mathematically, the high-temperature limit is seen when $e^{-\alpha}e^{\beta\varepsilon}$ is much larger than $\pm 1$ so that cases 2 and 3 above,

$$n_i = \frac{1}{e^{-\alpha}e^{\beta\varepsilon} \pm 1},$$

tend to case 1

$$n_i = \frac{1}{e^{-\alpha}e^{\beta\varepsilon}} = e^{\alpha}e^{-\beta\varepsilon}, \qquad (7.7.1)$$

which is the Boltzmann law where $n_i$ is understood to be the most probable occupation number.

Now we need to look more carefully at how these three distributions come about. We shall see that they arise essentially from counting differences, and that the counting differences arise from different restrictions on the particles in the case 2 and case 3 distributions. We shall be seeking the number of ways particles can be arranged to arrive at a given *configuration* (see below). All arrangements being equally probable, the configuration that can be formed in the greatest number of ways is the *most probable configuration*.

## 7.8 Permutations and Combinations

If we pick $N$ *distinguishable* objects and put them in an ordered arrangement, there are $N$ ways of picking the first object, $(N - 1)$ ways of picking the next, and so on, down to the last possible choice. This gives $N$ factorial (denoted $N!$) arrangements

$$^{N}P_N = N(N - 1)(N - 2) \cdots 1 = N! \qquad (7.8.1)$$

These are the *permutations* of $N$ objects taken $N$ at a time, denoted $^{N}P_N$.

An example is the number of permutations of particles $a$, $b$, and $c$ in three ordered levels, low, mid, and high, one particle to each level, books on shelves perhaps, but at this point, *not* energy levels. That comes later. The formula gives $3! = 3 \cdot 2 \cdot 1 = 6$ arrangements which we show in vertical columns of three:

| high | $c$ | $b$ | $c$ | $a$ | $b$ | $a$ |
| mid | $b$ | $c$ | $a$ | $c$ | $a$ | $b$ |
| low | $a$ | $a$ | $b$ | $b$ | $c$ | $c.$ |

If we have only two levels available, we have to put two choices in the same level. If we *cannot distinguish* between particles in the same level, we have reduced the number of arrangements by 1/2!. The number of arrangements we can construct under this restriction has been reduced to 3.

We can illustrate this by putting two particles in a "box" and identifying the particle outside the box, but only showing the box that contains the other two. We cannot give the vertical arrangement of the other two particles because we cannot "see" into the box:

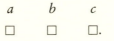

Arrangements $N$ of objects in groups of $n$, making no distinction within the groups, are called *combinations* of $N$ objects $n$ at a time, written $^N C_n$. For three objects arranged in two groups with distinction between the groups but not within a group,

$$^3 C_2 = \frac{3!}{2! \cdot 1!} = 3. \tag{7.8.2}$$

For four particles arranged in two groups of two, we have

$$^4 C_2 = \frac{4!}{2! \cdot 2!} = \frac{24}{2! \cdot 2!} = 6. \tag{7.8.3}$$

For $N$ particles arranged in groups of $n$,

$$^N C_n = \frac{N!}{\prod_i n_i!}. \tag{7.8.4}$$

## 7.9 Configurations

A distribution of $N$ particles over $k$ states, each containing some number of particles, $n_i$, where the states are distinguishable from one another but particles within a state are not, is a *configuration*, $n_0, n_1, n_2, \ldots, n_k$. The $n_i$ need not, and in general will not, be the same. Each set of

$n_0, n_1, n_2, \ldots, n_k, n'_0, n'_1, n'_2, \ldots, n'_k$, and so on, where at least one of the $n_i \neq n'_i$, is a different configuration. Each configuration has a *statistical weight* $W$ according to the number of ways it can be made up. For example,

are three ways of making up the configuration of indistinguishable particles with one particle in the upper state and two in the lower state. The configuration has a statistical weight

$$W = \frac{3!}{2!1!} = 3.$$

In general,

$$W = \frac{N!}{n_0! n_1! n_2! \cdots}, \quad W' = \frac{N!}{n'_0! n'_1! n'_2! \cdots}, \qquad (7.9.1)$$

and so on, which is also written

$$W = N! \prod_{i=0}^{k} \frac{1}{n_i!}. \qquad (7.9.2)$$

We are interested in situations in which a large but finite number of particles are distributed over levels that may be infinite in number. For any configuration, $n_0, n_1, n_2, \ldots, n_k$, the statistical weight of finding that configuration among all others is also the number of ways that configuration can be formed. An important assumption is that all arrangements are equally probable. That is why the probability of attaining a distribution is proportional to the number of ways it can be attained. If we know $W$ for all configurations, the probability of finding any one configuration is equal to its statistical weight divided by the statistical weights of all configurations

$$\text{prob} = \frac{W_i}{\sum_i W_i} \qquad (7.9.3)$$

Naturally, the largest $W$ leads to the highest probability.

A distribution over *energy* levels exists because energy can be passed from one particle to another. Thus we envision a distribution as a dynamic thing, with individual particles randomly exchanging places and occasionally seeking a new energy level. This means that the ratio $n_i/n_j$ is a ratio of the most probable values for $n_i$ and $n_j$. For large numbers of particles, however, fluctuations away from the most probable value are very rare. For a mole of particles (an immense number), one configuration so dominates the others that we can say that the system never deviates from it. That is why every time we measure a physical property of a macroscopic sample, we get the same answer. We do not find the sample in one configuration today and another configuration tomorrow, even though the individual particles are exchanging energy (changing place in the energy manifold) all the time.

Only a finite number of energy levels can be occupied by a finite (though large) number of particles. Occupied levels need not be contiguous. They may be separated by unoccupied levels. The presence of empty energy levels shows that zero factorial, $0! = 1$, is a reasonable definition, otherwise a configuration with an empty group ($n_i = 0$) would be undefined.

## 7.10 Stirling's Approximation

In examining statistical weights, we shall need Stirling's approximation for the logarithms of large factorials

$$\ln x! \cong x \ln x - x. \qquad (7.10.1)$$

Stirling's approximation is very good, even for $x$ as small as 100.

## 7.11 Constraints

We would like to find the dominant configuration for a number of particles distributed over an energy level spectrum, that is, we would like to find the configuration with the largest statistical weight. $W$ is a function of all the $n_i$, $W(n_0, n_1, n_2, \dots)$ so, if we set all the $\partial W / \partial n_i$ simultaneously equal to zero, we should have the configuration of maximum statistical weight. This does not use all the information we have; hence it will yield a less informative output than we might have. We can make use of some more of the information we have about the system with the help of a technique developed in the eighteenth century by J.-L. Lagrange. In general, if we can impose conditions upon a purely mathematical solution to a problem, where those conditions arise from the physics of the problem, we shall get a solution that is more meaningful (in a physical sense) than the original solution was. Stated in this way, the principle seems obvious, but it took an intellect like Lagrange's to show us how to do it.

From the way the problem has been set up,

$$N = \sum_i n_i, \tag{7.11.1}$$

because the total number of particles is the sum of the particles in each group, and

$$E = \sum_i n_i \varepsilon_i, \tag{7.11.2}$$

because the total energy is the energy at each level times the number of particles residing at that level. Imposing these two constraints on the problem of finding $W_{max}$ is done by the *Lagrangian method of undetermined multipliers*.

## 7.12 The Classical Boltzmann Distribution

Let us try out the Lagrangian method and Stirling's approximation on a problem for which we already know the answer: the Boltzmann distribution over nondegenerate energy levels of the harmonic oscillator. We have $W(n_0, n_1, n_2, \dots)$ and we have the rule that different ways of achieving configurations over a given energy-level spectrum are equally probable, so we wish to find the maximum *number of ways* a configuration can be constructed in order to find the most likely configuration. The maximum of a function $W = f(n_i)$ is found by setting all the $\partial W / \partial n_i = 0$ simultaneously (subject to constraints).

Finding $\partial \ln W / \partial n_i = 0$ achieves the same thing because both $W$ and $\ln W$ are monotonic functions of $n_i$. From $W = N! \prod_i (1/n_i!)$, we have

$$\ln W = \ln N! - \sum_i \ln n_i! \tag{7.12.1}$$

The conditions of constraint are $N = \sum_i n_i$ and $E = \sum_i n_i \varepsilon_i$. Both $N$ and $E$ are constant; hence

$$\delta N = \sum_i \delta n_i = 0 \quad \text{and} \quad \delta E = \sum_i \varepsilon_i \delta n_i = 0. \tag{7.12.2}$$

We have many independent variables and two conditions of constraint. By the Lagrangian method, this leads to two *undetermined multipliers*, call them $\alpha$ and $-\beta$, where the sign is arbitrary and is assigned according to convention.

A composite function $u$ that contains all three conditions must be maximized with respect to all the $n_i$. In differentiation over various arrangements to find the most probable configuration, we set a small variation in the composite function equal to zero,

$$\delta u = \delta \ln W + \alpha \sum_i \delta n_i - \beta \sum_i \varepsilon_i \delta n_i = 0. \tag{7.12.3}$$

We recognize that $\ln W$ is a function of all the $n_i$; hence

$$\delta \ln W = \frac{\partial \ln W}{\partial n_1} \delta n_1 + \frac{\partial \ln W}{\partial n_2} \delta n_2 + \cdots = \sum \frac{\partial \ln W}{\partial n_i} \delta n_i. \tag{7.12.4}$$

The differential of the composite function is

$$\delta u = \sum \frac{\partial \ln W}{\partial n_i} \delta n_i + \alpha \sum_i \delta n_i - \beta \varepsilon_i \sum_i \delta n_i = 0 \tag{7.12.5}$$

or

$$\sum \left( \frac{\partial \ln W}{\partial n_i} + \alpha - \beta \varepsilon_i \right) \delta n_i = 0. \tag{7.12.6}$$

We cannot expect the terms in this sum to exactly balance out to zero for any and every arbitrary $\delta n_i$; hence we are driven to the conclusion that at equilibrium each individual coefficient in the sum must be zero,

$$\frac{\partial \ln W}{\partial n_i} + \alpha - \beta \varepsilon_i = 0 \tag{7.12.7}$$

for the most probable configuration $W_{max}$.

Applying Stirling's approximation to the first term in equation 7.12.7,

$$\ln W = \ln N! - \sum \ln n_i!, \tag{7.12.8}$$

we get

$$\ln W = \ln N! - \sum n_i \ln n_i - n_i, \tag{7.12.9}$$

where we have not applied Stirling's approximation to $\ln N!$ ($N! = $ constant) because it will drop out in the differentiation, that is, $\frac{\partial \ln N!}{\partial n_i} = 0$; therefore

$$\frac{\partial \ln W}{\partial n_i} = \frac{\partial \ln N!}{\partial n_i} - \frac{\partial \sum n_i \ln n_i - n_i}{\partial n_i} = - \left[ n_i \frac{\partial \ln n_i}{\partial n_i} + \ln n_i \left( \frac{\partial n_i}{\partial n_i} \right) - \frac{\partial n_i}{\partial n_i} \right]$$

$$= -(1 + \ln n_i - 1) = - \ln n_i \tag{7.12.10}$$

because

$$n_i \frac{\partial \ln n_i}{\partial n_i} = n_i \frac{\left(\dfrac{\partial n_i}{n_i}\right)}{\partial n_i} = \frac{\partial n_i}{\partial n_i} = 1$$

and all other terms in the sum involving $\partial n_i / \partial n_j$ are zero. Filling in the conditions of constraint

$$-\ln n_i + \alpha - \beta \varepsilon_i = 0,$$
$$\ln n_i = \alpha - \beta \varepsilon_i \qquad\qquad (7.12.11)$$

and taking antilogarithms,

$$n_i = e^{\alpha - \beta \varepsilon_i}, \qquad\qquad (7.12.12)$$

which is the classical Boltzmann distribution, equation 7.7.1. If we set $\varepsilon_0 = 0$, we get $n_0 = e^{\alpha}$, whence

$$\frac{n_i}{n_0} = e^{-\beta \varepsilon_i}, \qquad\qquad (7.12.13)$$

which is equation 7.4.2.

## 7.13 Fine Structure

If there is some fine structure to the energy states that we have chosen for our derivation, the counting problem is changed in a subtle way. By fine structure we mean that what appears upon coarse examination to be one energy level $\varepsilon_i$ is actually split into $\gamma_i$ close-lying levels. For example, let $\gamma_1 = 2$ at $\varepsilon_i$.

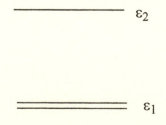

We have a two-level arrangement with a two-level fine structure and one particle in the lower level. There are two ways the fine structure can contribute to the arrangement:

$a$      $\cdots$

or

$$\cdots \qquad a.$$

The subtle point is this: both of these contributions from the fine structure constitute new ways of achieving the arrangement but they are not new arrangements. We have twice as many ways of achieving the same arrangement as we thought we had without fine structure:

$$W = N!\frac{2}{n_i!} \qquad (7.13.1)$$

instead of

$$W = N!\frac{1}{n_i!}. \qquad (7.13.2)$$

If we distribute two distinguishable particles over a fine structure of two levels, we get four ways (each given by a vertical column below) of arriving at the same arrangement:

$$
\begin{array}{cccc}
ab & a & b & \cdots \\
\cdots & b & a & ab.
\end{array}
$$

If we distribute three particles, we get eight ways of arriving at the same arrangement:

$$
\begin{array}{cccccccc}
abc & ab & ac & bc & a & b & c & \cdots \\
\cdots & c & b & a & bc & ac & ab & abc,
\end{array}
$$

and so on.

The sequence is $2, 4, 8, \ldots$, which we recognize as $2^n$ where $n$ is the number of particles distributed over a fine structure with two levels. If we play the same game with three levels, we find that we have to multiply the number of ways of achieving an arrangement without fine structure by $3^n$ to get the ways of achieving it with a fine structure of three. In general, we multiply by $\gamma_i^{n_i}$ where $\gamma_i$ is the number of levels in the fine structure of the $i$th arrangement and $n_i$ is the number of particles distributed over that arrangement. For many arrangements, we take the product of all possible arrangements just as we did without consideration of fine structure, that is,

$$W = \frac{N!}{n_0!n_1!n_2!\cdots} = N!\prod_i \frac{1}{n_i!} \qquad (7.13.3)$$

becomes

$$W = N! \frac{\gamma_0^{n_0} \gamma_1^{n_1} \gamma_2^{n_2}}{n_0! \, n_1! \, n_2!} \cdots = N! \prod_i \frac{\gamma_i^{n_i}}{n_i!}. \tag{7.13.4}$$

It should come as no surprise that, as the energy separation between fine structure levels approaches zero, we describe the levels as *degenerate*.

## 7.14 The Classical Case: A More General Derivation

Now that we are experts in the Lagrangian method of undetermined multipliers, we can obtain a more general derivation for the classical Boltzmann distribution. From

$$W = N! \prod_i \frac{\gamma_i^{n_i}}{n_i!} \tag{7.14.1}$$

we have

$$\ln W = \ln N! + \sum_i n_i \ln \gamma_i - \sum_i \ln n_i! \tag{7.14.2}$$

To find the maximum, let $n_i$ change by $\delta n_i$ under the constraints $\delta N = \sum \delta n_i = 0$ and $\delta E = \sum \varepsilon_i \delta n_i = 0$,

$$\delta \ln W + \alpha \sum_i \delta n_i - \beta \varepsilon_i \sum_i \delta n_i = 0, \tag{7.14.3}$$

where $\alpha$ and $-\beta$ are the undetermined multipliers. By Stirling's approximation,

$$\ln W = \ln N! + \sum_i (n_i \ln \gamma_i - n_i \ln n_i + n_i). \tag{7.14.4}$$

For a small variation $\delta n_i$,

$$\sum_i \left( \delta n_i \ln \gamma_i - \delta n_i \ln n_i - n_i \frac{\delta n_i}{n_i} + \delta n_i + \alpha \delta n_i - \beta \varepsilon_i \delta n_i \right) = 0$$

gives

$$\sum_i \left( \ln \gamma_i - \ln n_i + \alpha - \beta \varepsilon_i \right) \delta n_i = 0. \tag{7.14.5}$$

because $\delta \ln N! = 0$, $\delta \ln \gamma_i = 0$, and $-n_i \delta n_i / n_i + \delta n_i = 0$.

The coefficients of the sum must be zero:

$$\ln \gamma_i - \ln n_i + \alpha - \beta \varepsilon_i = 0,$$

$$-\ln \frac{n_i}{\gamma_i} + \alpha - \beta \varepsilon_i = 0. \tag{7.14.6}$$

Taking antilogarithms,

$$\ln \frac{n_i}{\gamma_i} = \alpha - \beta \varepsilon_i,$$

$$n_i = \gamma_i e^{\alpha - \beta \varepsilon_i}. \tag{7.14.7}$$

We could have said, by a somewhat circular definition of degeneracy, that $n_i = e^{\alpha - \beta \varepsilon_i}$ is the distribution function for nondegenerate levels and hence $n_i = \gamma_i e^{\alpha - \beta \varepsilon_i}$ must be the distribution function for a system in which the $i$th level has a degeneracy of $\gamma_i$.

## 7.15 Fermi-Dirac Counting

$$\gamma_i = 2.$$

The equation for the number of ways a configuration can be constructed from fermions can be built up in stepwise fashion. Suppose we start with a degeneracy of two and no particles. There is only one way this case can appear:

Ways

. . .        . . .                    1.

Now we add a particle. This can be done in two ways,

$a$      . . .

or

. . .      $a$.                    2.

Now add another particle *indistinguishable* from the first, remembering that fermions do not double up. This can be done in only one way,

$a$      $a$.                    1.

For fermions, there can never be more particles than states, $n_i \geq \gamma_i$, in this case 2. Our simple starting pattern is $1, 2, 1$.

$\gamma_i = 3.$

Now take a degeneracy of $\gamma_i = 3$. No particles gives

$$\cdots \qquad \cdots \qquad \cdots \hspace{5cm} 1.$$

Arbitrarily choose a first state. One particle either goes into the first state or it does not. If it does,

$$a \qquad \cdots \qquad \cdots \hspace{5cm} 1.$$

If it does not, we have an empty level plus a problem of one particle in two states, which is a problem we have already solved:

$$\cdots \qquad \text{(already solved)} \hspace{4cm} 2.$$

$$\text{total } 1 + 2 = 3.$$

Taking two indistinguishable particles, either the first particle selected goes into the first state or it does not. If it does, we have the problem of one particle in two levels,

$$a \qquad \text{(already solved)} \hspace{4cm} 2.$$

If it does not, we have

$$\cdots \qquad a \qquad a \hspace{5cm} 1$$

$$\text{total } 2 + 1 = 3.$$

Finally, for this degeneracy, take three particles

$$a \qquad a \qquad a \qquad 1.$$

The pattern is $1, 3, 3, 1$.

$\gamma_i = 4.$

For four degenerate states, we have

| | | | | | |
|---|---|---|---|---|---|
| $n = 0$ | $\cdots$ | $\cdots$ | $\cdots$ | $\cdots$ | 1 |
| $n = 1$ | $a$ | $\cdots$ | $\cdots$ | $\cdots$ | 1 |
| | $\cdots$ | | (already solved) | | 3 |

| $n = 2$ | $a$ | | (already solved) | | | 3 |
| | $\cdots$ | | (already solved) | | | 3 |
| $n = 3$ | $a$ | | (already solved) | | | 3 |
| | $\cdots$ | $a$ | $a$ | $a$ | | 1 |
| $n = 4$ | $a$ | $a$ | $a$ | $a$ | | 1. |

With $\gamma_i = 4, n_i = 0, 1, 2, 3, 4$ gives $1, 4, 6, 4, 1$. It is noteworthy that, with fermions, increasing $n_i$ at fixed $\gamma_i$ does not necessarily give more ways to construct a configuration. For example, going from three to four particles causes a decrease from four to one in the number of ways the arrangement can be achieved. The maximum number of arrangements comes at $n_i = 2$. The sequence with $\gamma_i = 1, 2, 3, 4$ combined with all possible values of $n_i$ gives $\{1\}, \{1 + 2 + 1\}, \{1 + 3 + 3 + 1\}, \{1 + 4 + 6 + 4 + 1\} = 1, 4, 8, 16$.

For a general expression of the number of ways fermions can contribute to a configuration, we expect $\gamma_i$ to go into the numerator of the fraction making up $W$. The exponent $n_i$ in $\gamma_i^{n_i}$ that we saw in the classical case arose through multiple occupancy of states, so we do not expect to see it here. We know that any value $(\gamma_i - n_i) < 0$ will not occur, and we also know that the restriction that fermions never occupy the same state will cause the number of ways of making up fermion configurations to be smaller than the number of ways classical particle configurations can be built up. Taking these factors together, a plausible modification of the classical distribution without the $n_i$ exponent and with a factor $(\gamma_i - n_i)!$ in the denominator would be to replace

$$W_{\text{classical}} = N! \prod_{i=1}^{\infty} \frac{\gamma_i^{n_i}}{n_i!} \tag{7.15.1}$$

with a product in which the individual terms are

$$W_{\text{Fermi-Dirac}} \equiv W_F = \frac{\gamma_i!}{n_i!(\gamma_i - n_i)!}. \tag{7.15.2}$$

We can see by inspection that this function goes through a maximum because $n_i!(\gamma_i - n_i)!$ is a maximum at either end of the range of $n_i$, causing a minimum in the denominator at midrange. A minimum in the denominator brings about a maximum in the fraction.

We can verify this function for $\gamma_i = 4$ over the values $n_i = 0, 1, 2, 3, 4$. We get

$$W_F = \frac{\gamma_i!}{n_i!(\gamma_i - n_i)!} = \frac{4!}{0!4!} = 1, \frac{4!}{1!3!} = 4, \frac{4!}{2!2!} = 6, \frac{4!}{3!1!} = 4, \frac{4!}{4!0!} = 1.$$

The series is $1, 4, 6, 4, 1 = 16$, as we expected from our Fermi counting procedure. Playing the Fermi counting game with larger degeneracies is simple but tedious. In the end, we need to take the product of all the $W_F$ just as we did in the classical case. We arrive at

$$W_F = \prod_{i=1}^{\infty} \frac{\gamma_i!}{n_i!(\gamma_i - n_i)!} \tag{7.15.3}$$

as the number of ways a configuration can be constructed from Fermi-Dirac particles.

## 7.16 The Fermi-Dirac Distribution Function

Applying the Lagrangian method of undetermined multipliers, let

$$\delta \ln W + \alpha \sum_i \delta n_i + \beta \varepsilon_i \sum_i \delta n_i = 0. \tag{7.16.1}$$

For fermions,

$$\ln W_F = \sum_i \ln \gamma_i! - \ln n_i! - \ln (\gamma_i - n_i)! \tag{7.16.2}$$

By Stirling's approximation (equation 7.10.1),

$$\ln W_F = \sum_i \gamma_i \ln \gamma_i - \gamma_i - n_i \ln n_i + n_i - (\gamma_i - n_i) \ln (\gamma_i - n_i) + (\gamma_i - n_i), \tag{7.16.3}$$

and for a small change $\delta n_i$,

$$\delta \ln W_F = \left[ \sum_i 0 - 0 + (-1 - \ln n_i + 1) + (1 + \ln (\gamma_i - n_i) - 1) \right] \delta n_i. \tag{7.16.4}$$

Simplifying and putting in the conditions of constraint,

$$\delta \ln W_F = \sum_i (- \ln n_i + \ln (\gamma_i - n_i) + \alpha - \beta \varepsilon_i) \delta n_i = 0. \tag{7.16.5}$$

The individual coefficients of $\delta n_i$ are zero,

$$- \ln n_i + \ln (\gamma_i - n_i) + \alpha - \beta \varepsilon_i = 0,$$

$$\ln \frac{(\gamma_i - n_i)}{n_i} + \alpha - \beta\varepsilon_i = 0,$$

$$\ln \frac{(\gamma_i - n_i)}{n_i} = \ln \left(\frac{\gamma_i}{n_i} - 1\right) = -(\alpha - \beta\varepsilon_i). \qquad (7.16.6)$$

Taking antilogarithms,

$$\left(\frac{\gamma_i}{n_i} - 1\right) = e^{-(\alpha-\beta\varepsilon_i)},$$

$$\frac{\gamma_i}{n_i} = e^{-\alpha+\beta\varepsilon_i} + 1, \qquad (7.16.7)$$

or

$$\frac{n_i}{\gamma_i} = \frac{1}{e^{-\alpha+\beta\varepsilon_i} + 1} \qquad (7.16.8)$$

for a distribution of fermions with $\gamma_i$ degeneracy. In the absence of degeneracy, we have

$$n_i = \frac{1}{e^{-\alpha+\beta\varepsilon_i} + 1} \qquad (7.16.9)$$

which is equivalent to equation 7.7.2 (case 2, section 7.7).

## 7.17 Bose-Einstein Counting

In Bose-Einstein counting, we wish to find out how many ways $n_i$ indistinguishable particles can be arranged in an energy level $\varepsilon_i$ without any restriction on the number of particles at that level. Since there is no prescribed occupation number, a level is never "filled." For example, at a level with $\gamma_i = 2$, going from 0 to 5 particles we have the following:

$\gamma_i = 2.$

|  |  |  | Ways |
|---|---|---|---|
| $n = 0$ | . . . | . . . | 1 |
| $n = 1$ | $a$ | . . . |  |
|  | . . . | $a$ | 2 |

| | | | |
|---|---|---|---|
| $n = 2$ | ... | $aa$ | |
| | $a$ | $a$ | |
| | $aa$ | ... | 3 |
| $n = 3$ | ... | $aaa$ | |
| | $a$ | $aa$ | |
| | $aa$ | $a$ | |
| | $aaa$ | ... | 4 |
| $n = 4, 5, \text{etc.}$ | | | 5, 6, etc. |

At a level with $\gamma_i = 3$, going from 0 to 3 particles gives the following:

$\gamma_i = 3.$

| | | | | Ways |
|---|---|---|---|---|
| $n = 0$ | ... | ... | ... | 1 |
| $n = 1$ | $a$ | ... | ... | |
| | ... | $a$ | ... | |
| | ... | ... | $a$ | 3 |
| $n = 2$ | $aa$ | ... | ... | |
| | ... | $aa$ | ... | |
| | ... | ... | $aa$ | |
| | $a$ | $a$ | ... | |
| | $a$ | ... | $a$ | |
| | ... | $a$ | $a$ | 6 |
| $n = 3$ | $aaa$ | ... | ... | |
| | ... | $aaa$ | ... | |
| | ... | ... | $aaa$ | |
| | $aa$ | $a$ | ... | |
| | $aa$ | ... | $a$ | |
| | $a$ | $aa$ | ... | |
| | ... | $aa$ | $a$ | |
| | $a$ | ... | $aa$ | |
| | ... | $a$ | $aa$ | |
| | $a$ | $a$ | $a$ | 10. |

From our series $1, 3, 6, 10$, we can look for a regularity and speculate that it will be followed further in the series. Notice that

$$0 + 1 = 1,$$
$$0 + 1 + 2 = 3,$$
$$0 + 1 + 2 + 3 = 6,$$
$$0 + 1 + 2 + 3 + 4 = 10.$$

We have recreated the series $1, 3, 6, 10$ for the first four members of the filling of a threefold degenerate level. Rather than go through the counting routine again, let us see in the next section whether the fifth member in the series will be

$$0 + 1 + 2 + 3 + 4 + 5 = 15.$$

## 7.18 The Bose-Einstein Statistical Weights $W_B$

To generalize this kind of counting, suppose we have $n_i$ particles separated by minute energy barriers. Let us take $n_i = 3$ particles occupying three intervals separated by two barriers,

$$a \mid a \mid a.$$

The number of permutations of particles *and* barriers is $n_i + \gamma_i - 1 = 5!$ If we take $5!$ as the number of arrangements that can contribute to a configuration, we have overcounted by the number of permutations of identical particles because exchange of an $a$ for an $a$ gives the same arrangement as before the exchange. This means that we must divide the total number of permutations by $n_i = 3!$ to compensate for the particle overcount. We have also overcounted by not recognizing that the two barriers can be exchanged without altering the arrangement. To compensate for this overcount, we must divide by the number of barriers $2!$ The number of barriers will always be one less than the number of intervals they separate, that is, $(\gamma_i - 1)$ no matter what the number $\gamma_i$. The number of permutations divided by the particle overcount and the barrier overcount is

$$W_{\text{Bose-Einstein}} \equiv W_B \underset{?}{=} \frac{(n_i + \gamma_i - 1)!}{n_i!(\gamma_i - 1)!} \tag{7.18.1}$$

We set this down as a plausible test equation for finding one configuration of Bose-Einstein particles (bosons). For the counting problem of three particles in a level with a degeneracy of 3, $a \mid a \mid a$,

$$W_B = \frac{(n_i + \gamma_i - 1)!}{n_i!(\gamma_i! - 1)!} = \frac{(3+2)!}{3!2!} = \frac{120}{12} = 10 \qquad (7.18.2)$$

in agreement with our prior count. Now, to test our speculation as to the result of the series

$$0 + 1 + 2 + 3 + 4 + 5 = 15$$

as the number of arrangements of four bosons in three degenerate levels, we set

$$W_B = \frac{(n_i + \gamma_i - 1)!}{n_i!(\gamma_i - 1)!} = \frac{(4+2)!}{4!2!} = \frac{720}{48} = 15.$$

The equation and the series agree. Many calculations of this kind will verify that the equation

$$W_B = \frac{(n_i + \gamma_i - 1)!}{n_i!(\gamma_i - 1)!} \qquad (7.18.3)$$

is in fact the number of boson arrangements for the $i$th energy level. For all arrangements over all levels, as before, we take the product

$$W_B = \prod_i \frac{(n_i + \gamma_i - 1)!}{n_i!\,(\gamma_i - 1)!}. \qquad (7.18.4)$$

## 7.19 The Bose-Einstein Distribution Function

Applying the Lagrangian method of undetermined multipliers to boson arrangements (equation 7.18.4) gives

$$\frac{n_i}{\gamma_i} = \frac{1}{e^{-\alpha + \beta \varepsilon_i} - 1} \qquad (7.19.1)$$

for a distribution of bosons with $\gamma_i$ degeneracy (see the problems at the end of this chapter). In the absence of degeneracy, we have

$$n_i = \frac{1}{e^{-\alpha + \beta \varepsilon_i} - 1} \qquad (7.19.2)$$

which is equivalent to equation 7.7.3 (case 3, section 7.7).

## 7.20 Summary Equations

A convenient form for particle distributions is

$$\frac{n_i}{\gamma_i} = (e^{-\alpha + \beta \varepsilon_i} + \delta)^{-1},$$ (7.20.1)

where $\delta = 0, 1,$ or $-1$ for Boltzmann, Fermi, and Bose distributions, respectively.

Since $\alpha$ is an undetermined multiplier, we can change its mathematical form in an equilibrium (constant-temperature) distribution to $\varepsilon/k_B T$ without physical consequence. We already know that $\beta = 1/k_B T$. For some applications, the summary form

$$\frac{n_i}{\gamma_i} = \left(e^{-\alpha + \beta \varepsilon_i} + \delta\right)^{-1} = \left(e^{-\varepsilon/k_B T + \varepsilon_i/k_B T} + \delta\right)^{-1}$$

$$= \left(e^{(\varepsilon_i - \varepsilon)/k_B T} + \delta\right)^{-1}$$ (7.20.2)

is useful. The distributions are

$$n_i(\varepsilon) = n_0 e^{-\varepsilon_i/k_B T} \quad \text{(Boltzmann)},$$ (7.20.3)

$$F(\varepsilon) = \left(e^{(\varepsilon_i - \varepsilon_F)/k_B T} + 1\right)^{-1} \quad \text{(Fermi-Dirac)},$$ (7.20.4)

$$B(\varepsilon) = \left(e^{(\varepsilon_i - \varepsilon_B)/k_B T} - 1\right)^{-1} \text{(Bose-Einstein)}.$$ (7.20.5)

## 7.21 An Alternative Derivation for Fermions and Bosons

Consider the process in which a *system* is in contact with a thermal bath at temperature $T$. The system has an empty quantum state $Q_i$ which acquires a particle

$$Q_i(0) \rightarrow Q_i(1).$$

The particle has to come from somewhere in the system, absorbing $\varepsilon_i - \bar{\varepsilon}$ of energy, where $\bar{\varepsilon}$ is the energy of the particles selected from anywhere in the system except $\varepsilon_i$. The average energy of the particles over many repetitions of this process is the total energy divided by the number of particles, $\bar{\varepsilon} = E/N$.

The entropy decrease of the bath is

$$\Delta S_{\text{bath}} = -\frac{\varepsilon_i - \bar{\varepsilon}}{T} \tag{7.21.1}$$

for one particle. Since the particle comes from any one of many possible states in the system and ends up in one specific state, the process $Q_i(0) \rightarrow Q_i(1)$ is an ordering process and is accompanied by an entropy decrease in the system. The average entropy decrease of the system over transfer of many particles is the system entropy divided by $N$,

$$\Delta S_{\text{system}} = -\frac{S}{N} = -\bar{s}. \tag{7.21.2}$$

The total entropy change is

$$\Delta S_{\text{total}} = \Delta S_{\text{bath}} + \Delta S_{\text{system}}$$

$$= -\frac{\varepsilon_i - \bar{\varepsilon}}{T} - \bar{s} = -\frac{\varepsilon_i - (\bar{\varepsilon} - T\bar{s})}{T}. \tag{7.21.3}$$

Let us define $\bar{\varepsilon} - T\bar{s} \equiv \mu$ as a *chemical potential* so that

$$\Delta S_{\text{total}} = -\frac{\varepsilon_i - \mu}{T} = \frac{\mu - \varepsilon_i}{T}. \tag{7.21.4}$$

By Boltzmann's definition, entropy is a thermodynamic state function that is proportional to the logarithm of the number of ways a system can be made up,

$$S = k_B \ln W. \tag{7.21.5}$$

The probability of finding a system in a given state is proportional to the number of ways it can be made up,

$$\Delta S_{\text{total}} = k_B \ln \frac{p_i(1)}{p_i(0)} \tag{7.21.6}$$

for the process $Q_i(0) \rightarrow Q_i(1)$ where $p_i(n)$ is the probability of finding $n$ particles in state $i$. This leads to

$$\ln \frac{p_i(1)}{p_i(0)} = \frac{\Delta S_{\text{total}}}{k_B} = \frac{\mu - \varepsilon_i}{k_B T} \tag{7.21.7}$$

and

$$\frac{p_i(1)}{p_i(0)} = e^{(\mu - \varepsilon_i)/k_B T} = e^{\mu/k_B T} e^{-\varepsilon_i/k_B T}. \tag{7.21.8}$$

The notation $\lambda = e^{\mu/k_B T}$ is often used. In this notation,

$$\frac{p_i(1)}{p_i(0)} = \lambda e^{-\varepsilon_i/k_B T}. \tag{7.21.9}$$

The quantity $\lambda$ is called the *absolute activity* of a particle in the system.

## 7.22  Fermions (Again)

For fermions, no more than one particle can occupy a state so the probability

$$p_i(n) = 0 \quad \text{if } n \geq 2. \tag{7.22.1}$$

The average number of particles $\bar{n}_i$ in state $i$ is

$$\bar{n}_i = \frac{\sum n p_i(n)}{\sum p_i(n)} = \frac{0 \times p_i(0) + 1 \times p_i(1)}{p_i(0) + p_i(1)}$$

$$= \frac{1}{\dfrac{p_i(0)}{p_i(1)} + 1} = \frac{1}{\lambda^{-1} e^{\varepsilon_i/k_B T} + 1}$$

$$= \frac{1}{e^{(\varepsilon_i - \mu)/k_B T} + 1} \tag{7.22.2}$$

because $\lambda^{-1} = e^{-\mu/k_B T}$. This can be written

$$\bar{n}_i = \left(e^{(\varepsilon_i - \mu)/k_B T} + 1\right)^{-1} \tag{7.22.3}$$

where, by comparison to the summary equation for fermions, it is evident that $\mu = \varepsilon_F$.

## 7.23  Bosons (Again)

Because of the difference in restrictions on the number of particles allowed in a quantum state, we consider the process

$$Q_i(0) \rightarrow Q_i(1)$$

where $p_i(n) \neq 0$ if $n \geq 2$. Now the entropy changes for one particle in the fermion case become changes for $n$ particles in the boson case:

$$\Delta S_{\text{bath}} = -n \frac{\varepsilon_i - \bar{\varepsilon}}{T} \tag{7.23.1}$$

and

$$\Delta S_{\text{system}} = -n\frac{S}{N} = -n\bar{s},\qquad(7.23.2)$$

where

$$\Delta S_{\text{total}} = \Delta S_{\text{bath}} + \Delta S_{\text{system}}$$

$$= -n\frac{\varepsilon_i - \bar{\varepsilon}}{T} - n\bar{s} = -n\left(\frac{\varepsilon_i - (\bar{\varepsilon} - T\bar{s})}{T}\right)$$

$$= -n\left(\frac{\varepsilon_i - \mu}{T}\right).\qquad(7.23.3)$$

There is only one way of arranging $n$ particles in quantum state $i$ because the particles are indistinguishable. (Exchanging identical particles within a state does not bring about a new arrangement.) From Boltzmann's equation $S = k_B \ln W$,

$$\Delta S_{\text{total}} = k_B \ln \frac{p_i(n)}{p_i(0)} = -n\left(\frac{\varepsilon_i - \mu}{T}\right)\qquad(7.23.4)$$

and

$$\frac{p_i(n)}{p_i(0)} = e^{n\mu/k_B T} e^{-n\varepsilon_i/k_B T} = \lambda^n e^{-n\varepsilon_i/k_B T}.\qquad(7.23.5)$$

The average population of bosons in state $i$ is

$$\bar{n}_i = \frac{\sum np_i(n)}{\sum p_i(n)} = \frac{\sum n\lambda^n e^{-n\varepsilon_i/k_B T}}{\sum \lambda^n e^{-n\varepsilon_i/k_B T}}\qquad(7.23.6)$$

where the $p_i(0)$ cancel. To evaluate this quotient, let $x = \lambda e^{-\varepsilon_i/k_B T}$, whence

$$\bar{n}_i = \frac{\sum nx^n}{\sum x^n}.\qquad(7.23.7)$$

We saw in section 4.5 that the sum $\sum_{n=0} x^n = 1 + x + x^2 + x^3 + \cdots$ is

$$\sum_{n=0} x^n = \frac{1}{(1-x)}\qquad(7.23.8)$$

(provided that $x$ is in the interval $-1 < x < 1$). Taking the derivative of the denominator in our quotient 7.23.7,

$$\frac{d}{dx}\sum_{n=0} x^n = \sum_{n=0} nx^{n-1}.$$

We can multiply by $x$ to get the numerator

$$x \frac{d}{dx} \sum_{n=0} x^n = \sum_{n=0} n x^n, \tag{7.23.9}$$

but we also know that

$$x \frac{d}{dx} \sum_{n=0} x^n = x \frac{d}{dx} \frac{1}{(1-x)} = \frac{x}{(1-x)^2}; \tag{7.23.10}$$

therefore, equating the last two results,

$$\sum_{n=0} n x^n = \frac{x}{(1-x)^2}. \tag{7.23.11}$$

Now we have everything we need to evaluate the quotient 7.23.7, which turns out to be quite simple:

$$\bar{n}_i = \frac{\sum n x^n}{\sum x^n} = \frac{\frac{x}{(1-x)^2}}{\frac{1}{(1-x)}} = \frac{x}{(1-x)}. \tag{7.23.12}$$

Expressing $x$ once again as $x = \lambda e^{-\varepsilon_i / k_B T}$, we get

$$\bar{n}_i = \frac{x}{(1-x)} = \frac{\lambda e^{-\varepsilon_i / k_B T}}{1 - \lambda e^{-\varepsilon_i / k_B T}} \tag{7.23.13}$$

$$= \frac{1}{\frac{1}{\lambda} e^{\varepsilon_i / k_B T} - 1} = \frac{1}{e^{(\varepsilon_i - \mu)/k_B T} - 1}$$

$$= \left( e^{(\varepsilon_i - \mu)/k_B T} - 1 \right)^{-1}. \tag{7.23.14}$$

This is equivalent to case 3, section 7.7.

## 7.24 Reduction to the Classical Case

If $\varepsilon_i \gg \mu$, the exponential $e^{(\varepsilon_i - \mu)/k_B T}$ is large compared to 1, and both the Fermi-Dirac and the Bose-Einstein distributions reduce to

$$\bar{n}_i = \frac{1}{e^{(\varepsilon_i - \mu)/k_B T} - 1} = \lambda e^{-\varepsilon_i / k_B T}, \tag{7.24.1}$$

which is Boltzmann's law for a harmonic oscillator, provided that we take the average population of a level $\bar{n}_i$ as the probable population of state $i$, and $\lambda = n_0$.

## 7.25 The Entropy

Up to this point, our use of the Boltzmann distribution has rested on the rather shaky foundation of the barometric equation (section 2.11) for decrease in air pressure with altitude, which might (with luck) be empirically confirmed on a quiet day. Now let us revisit this law and look at it in a more rigorous way.

Boltzmann, a believer in atomic-molecular theory long before it was universally accepted, proposed that "properties" of collections of atoms tend to go in a "spontaneous" way such that a certain function approaches an extremum (maximum or minimum). During many years of dispute, which became at times factional and even personal, Boltzmann was driven from some of his ideas concerning the function (which he called the $H$ function) but certain other concepts were strengthened and clarified by the necessity to defend them. Boltzmann's function evolved over time into the entropy $S$ which is *maximized* during a spontaneous change (see also the ingenious work of S. Carnot).

Boltzmann was at first reluctant to concede that the basis of the entropy is probabilistic, that the equilibrium state of a system is its most *probable* state, and that any physical property that we observe is the one corresponding to its most probable configuration. (The evident immutability of the laws of thermodynamics arises from the vanishingly small probability of any fluctuation or deviation from them.) Eventually Boltzmann came or was driven to the probabilistic point of view, much derided by some of his contemporaries, who believed that physics based on the laws of probability isn't really physics. Boltzmann's point of view is, however, the one we hold today.

Thermodynamic state functions are additive, therefore for states 1 and 2,

$$S_{1+2} = S_1 + S_2. \tag{7.25.1}$$

If we take the first *configuration* of state 1 in combination with *all configurations* of state 2, we get the configurations of the composite state $1 + 2$. If we take the second configuration of state 1 in combination with all configurations of state 2, we get another set of configurations of the composite state. If we keep on doing this, we finally get all configurations of state 1 combined with all configurations of state 2, which gives us all configurations of the composite state. The number of configurations in the composite state is the number of ways of configuring state 1, $W_1$ times the number of ways of configuring state 2, $W_2$. If two functions $S_1 = f(W_1)$ and $S_2 = f(W_2)$ are related such that the sum of $S_1$ and $S_2$ is proportional to the product of $W_1$ and $W_2$ then there is a logarithmic relation between them:

$$S_1 + S_2 \propto \ln W_1 W_2 \qquad (7.25.2)$$

or, in general,

$$S \propto \ln W, \qquad (7.25.3)$$

which we now write as an equality with a proportionality constant:

$$S = k_B \ln W \qquad (7.25.4)$$

The proportionality constant $k_B$ is now called Boltzmann's constant in honor of its discoverer.

Because the probability of a state is directly proportional to the number of configurations accessible to it,

$$\Delta S = S_2 - S_1 = k_B \left( \ln W_2 - \ln W_1 \right)$$
$$= k_B \ln \frac{W_2}{W_1} = k_B \ln \frac{p_2}{p_1} \qquad (7.25.5)$$

where $p_1$ and $p_2$ are the probabilities of states 1 and 2. This is just the argument that we used in section 7.21.

Suppose now that a system in quantum state $i$ moves to quantum state $j$. There is one configuration in quantum state $i$ and one configuration in quantum state $j$; therefore

$$\Delta S_{\text{system}} = 0. \qquad (7.25.6)$$

Assuming the system to be in thermal contact with a bath with which it can exchange energy,

$$\Delta S_{\text{bath}} = -\frac{\varepsilon_j - \varepsilon_i}{T}, \qquad (7.25.7)$$

because the energies of the quantum states will be different and the energy difference must be made up (supplied or absorbed) by the bath. The total entropy change is

$$\Delta S_{\text{total}} = -\frac{\varepsilon_j - \varepsilon_i}{T} \qquad (7.25.8)$$

but we already know that, by Boltzmann's principle,

$$\Delta S_{\text{total}} = k_B \ln \frac{W_j}{W_i} = k_B \ln \frac{p_j}{p_i}, \qquad (7.25.9)$$

where $W_j$ is the number of configurations of the combined system when it is in state $j$ and $W_i$ is the number of configurations of the combined system when it is in state $i$. Because the probabilities are proportional to $W$,

$$k_B \ln \frac{p_j}{p_i} = -\frac{\varepsilon_j - \varepsilon_i}{T},$$

$$\ln \frac{p_j}{p_i} = -\frac{\varepsilon_j - \varepsilon_i}{k_B T}, \tag{7.25.10}$$

and

$$\frac{p_j}{p_i} = e^{-(\varepsilon_j - \varepsilon_i)/k_B T}. \tag{7.25.11}$$

If we take our "system" to be a large number of particles, the ratio of probabilities is the ratio of the expected or most probable numbers of particles in state $j$ relative to state $i$; hence

$$\frac{n_j}{n_i} = e^{-(\varepsilon_j - \varepsilon_i)/k_B T}. \tag{7.25.12}$$

We can express this law as a proportionality with a proportionality constant $A$,

$$n_j = A e^{-\varepsilon_j/k_B T}. \tag{7.25.13}$$

We note that

$$\sum n_j = N = \sum A e^{-\varepsilon_j/k_B T} = A \sum e^{-\varepsilon_j/k_B T} \tag{7.25.14}$$

where $N$ is the number of particles. Solving for $A$,

$$A = \frac{N}{\sum\limits_{\text{states}} e^{-\varepsilon_j/k_B T}} = \frac{N}{Q}, \tag{7.25.15}$$

where we have used the definition of the *canonical partition function*

$$Q \equiv \sum_{\text{states}} e^{-\varepsilon_j/k_B T} \tag{7.25.16}$$

to obtain a general form of Boltzmann's principle (equation 2.11)

$$n_j = \frac{N}{Q} e^{-\varepsilon_j/k_B T}. \tag{7.25.17}$$

## 7.26 A Note from Classical Thermodynamics: The Fundamental Equation

We have discussed the probabilistic (nonclassical) definition of entropy but we have also used classical expressions of the form

$$\Delta S = \frac{\varepsilon_j - \varepsilon_i}{T} \qquad (7.26.1)$$

without justification. If we add the restriction that the process involving exchange of an amount of energy $\varepsilon_j - \varepsilon_i$ between a system and a bath or reservoir is a reversible exchange of heat, $q_{rev}$, we have

$$dS = \frac{dq_{rev}}{T}, \qquad (7.26.2)$$

which is a statement of the second law of thermodynamics within the framework of classical thermodynamics. A statement of the first law of thermodynamics can be made in the form

$$dU = dq + p\,dV \qquad (7.26.3)$$

for a process involving only heat exchange and pressure-volume work. We note that these laws are consistent with our statistical treatment. Because of this consistency, if the statistical approach is wrong, we might expect classical thermodynamics to be wrong as well, which does not appear to be the case. The two basic laws of thermodynamics can be stated in a combined form as

$$dU = T\,dS + p\,dV, \qquad (7.26.4)$$

which we shall call the *fundamental equation* of thermodynamics for a system that does not exchange matter with its surroundings.

### PROBLEMS

**7.1.** Show that

$$-\frac{dq}{d\beta} = \sum \varepsilon_i e^{-\beta \varepsilon_i}$$

as used in obtaining equation 7.5.3.

7.2. Three degenerate levels exist within a state. Three photons are distributed among them. How many ways can arrangements be found such that

   (a) One level has all three photons.
   (b) One level has two photons and one level has the other photon.
   (c) Each level has a photon.

   What is the probability of distributions $a, b$, and $c$ over all three levels?

7.3. (a) How many ways can two indistinguishable particles be distributed over a state with twofold fine structure?
   (b) What is the probability of finding at least one of two indistinguishable particles in the upper level? (The fine structure energies are sufficiently different that one can tell the occupation number of each but not so different as to bias the distribution.)
   (c) What is the probability of finding one particle, but not both, in the upper level?

7.4. Work problem 7.3 for two distinguishable particles distributed over a state with twofold fine structure.

7.5. (a) How many ways can three indistinguishable particles be distributed over a state with twofold fine structure?
   (b) What is the probability of finding at least one particle in the upper level?
   (c) What is the probability of finding exactly one particle in the upper level?

7.6. (a) How many ways can three indistinguishable particles be distributed over two levels with twofold degeneracy in the lower level?
   (b) What is the probability of finding at least one particle in the upper level?
   (c) What is the probability of finding exactly one particle in the upper level?

7.7. Apply the Lagrangian method of undetermined multipliers

$$\delta \ln W + \alpha \sum_i \delta n_i + \beta \varepsilon_i \sum_i \delta n_i = 0$$

to boson arrangements

$$W_B = \prod_i \frac{(n_i + \gamma_i - 1)!}{n_i! \, (\gamma_i - 1)!}$$

to find

$$\frac{n_i}{\gamma_i} = \frac{1}{e^{-\alpha + \beta \varepsilon_i} - 1}$$

for the distribution of bosons with $\gamma_i$ degeneracy.

# Eight

## Consequences of the Fermi-Dirac Distribution

The WORKING HYPOTHESIS for the photoelectric effect as described by Einstein implies a three dimensional "lattice" of positive metal atoms immersed in a sea of noninteracting electrons which are free to move anywhere within the lattice. This suggests that, for calculating the heat capacity, the model can be divided into a *lattice part* and an *electronic part*. The lattice part has already been treated in the sections on the Einstein equation and especially Debye equation, which treats a macroscopic crystal lattice vibrating harmonically in three dimensions. In this chapter, we shall look at the electronic part of the model.

### 8.1 The Electron Gas

For the electronic part, we might expect some resemblance to an "ideal gas" of electrons. The contribution of ideal monatomic gases to the molar heat capacity at constant volume is well known as $\frac{3}{2}R$ ($\frac{1}{2}R$ for each of three degrees of translational freedom) but a prediction that electronic translational motion will contribute $\frac{3}{2}R$ to the heat capacity of a metal is wrong. The electronic contribution to the heat capacity is not zero but it is so small that it can be detected only with difficulty at very low temperatures. We enquire why the failure of this prediction is so complete, why the electronic heat capacity bears no resemblance at all to the rough speculative model we have proposed. Must we throw out the entire model, or can we find some special characteristic of the Fermi electron gas that explains electronic heat capacity in metals but retains the model's essential characteristics?

### 8.2 The Fermi Sea

The most obvious difference between electrons and atoms or molecules is size. Electrons are much smaller than atoms or molecules, being roughly 2000 times less massive than the lightest atom, hydrogen. As a semiquantitative model, let us think of a cube of metal, having one molar mass, as

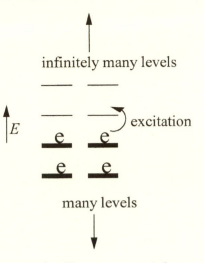

Figure 8.2.1. Highest occupied and lowest unoccupied energy levels of an electron gas.

a "box" within which electrons are free to move. By a calculation analogous to the "particle in a box" problem of elementary physical chemistry, energy levels can be found which are arranged in some sort of energy ladder as in figure 8.2.1. The lowest $\frac{1}{2}N_A$ energy levels in this model are occupied by electrons, where $N_A$ is the Avogadro number. There are $\frac{1}{2}N_A$ occupied energy levels because electrons, which are fermions, occupy twofold-degenerate energy levels in pairs, spin up and spin down. The levels above the lowest $\frac{1}{2}N_A$ levels are vacant.

The point is not the precise arrangement of energy levels, which probably bears little resemblance to the particle in a box model, but that there are very many filled energy levels below the first vacant energy level. No matter what their spacing, there is an Avogadro's number of electrons in the filled set of orbitals, an immense number. With each level, the energy of the next level must be higher than the level below it, so, whatever the arrangement, the energy of the highest occupied level must be very high indeed. The highest level is called the *Fermi energy* $\varepsilon_F$. Electrons at the level with energy $\varepsilon_F$ are said to be on the surface of the *Fermi sea*. Other electrons are at greater depths in the Fermi sea. The Fermi energy $\varepsilon_F = k_B T_F$ divided by $k_B$ has units of temperature $T_F = \varepsilon_F / k_B$. This temperature is called the *Fermi temperature*.

At 0 K, the surface of the Fermi sea is perfectly flat, but at a temperature slightly above 0 K, a thin cloud of electrons appears above the surface of the sea, consisting of electrons in dynamic equilibrium with excited states. There is a diminution of the density of the Fermi sea just below its surface

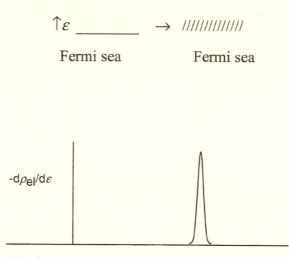

Figure 8.2.2. The sharp edge between the Fermi sea and unfilled orbitals (left) becomes indistinct at higher temperatures (right). The vertical line on the left is called a Dirac delta function. It becomes a Gaussian on the right.

owing to the vacant energy states left by the electrons that have gone into excited states. At the same time, the electron density just above the surface of the Fermi sea is augmented from zero to some nonzero value.

The electronic contribution to the heat capacity requires absorption of heat by the sample from the surroundings to bring about excitation of an electron to the first unoccupied energy state or higher.

At absolute zero, this process simply does not take place. Even at laboratory temperatures, the Gaussian in figure 8.2.2 is not very broad. Of all the electrons in the Fermi sea, only a few have absorbed thermal energy to go into an excited state.

The energetic difference between an ideal gas of atoms and an electron gas lies in the difference between the high energy at the surface of the electron sea and the small energy gaps between the translational levels of an ideal gas of atoms. Upon absorption of $\frac{3}{2}RT$ of thermal energy, where $T$ is small, a mole of atoms is excited into higher translational states, but only ~0.01 mol of electrons is excited into the next higher electronic state for a typical metal. A fixed increment $\Delta E_{small}$ is distributed over all of the atoms in the ideal gas, but only a few electrons at the surface of the Fermi

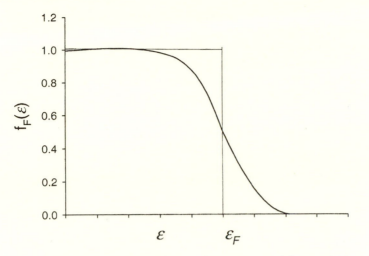

Figure 8.3.1. The Fermi-Dirac distribution $f_F(\varepsilon)$ at 0 K and $k_B T = 0.20\varepsilon_F$. The average number of electrons in an energy state is $\bar{n}_i = f_F(\varepsilon)$. The Fermi energy is $\varepsilon_F$.

sea participate in heat absorption. For a fixed $\Delta E_{\text{small}}$, the heat capacity must be large in the case of the ideal gas and small in the case of the electron gas.

## 8.3 The Fermi Distribution

At 0 K, an energy level in figure 8.2.1 is either occupied or it is not. The probability of finding an electron in an energy state is 1 for those states below the Fermi energy and 0 for those states above it. This is shown by the Dirac delta function in figure 8.2.2 and the step function with a vertical at $\varepsilon_F$ in figure 8.3.1. Figure 8.3.1 shows the Fermi distribution function $\bar{n}_i = f_F(\varepsilon)$, which gives the average or expected number of electrons at any energy $\varepsilon$ as a function of $\varepsilon$. The distribution is shown at two temperatures, 0 K and $0.20T_F$.

To measure heat capacity, we must supply a small amount of energy to the system. Near 0 K, most electrons do not absorb energy because they are deep within the Fermi sea. At higher temperatures, the band of thermal energy $k_B T$ is wider and goes beneath the surface of the Fermi sea. More electrons are excited above $\varepsilon_F$. This is seen in the reverse sigmoidal curve in figure 8.3.1, which shows that some energy states below $\varepsilon_F$ are empty and an equal number of states above $\varepsilon_F$ are filled. At temperatures below $T_F$, the curve is symmetrical about $f_F(\varepsilon) = \frac{1}{2}$ because the number of electrons lost below $\varepsilon_F$ is equal to the number gained above $\varepsilon_F$. At higher temperatures, the slope of the curve in figure 8.3.1 gradually becomes less steep because there are more electrons in excited states and more "holes" where electrons used to be at $\varepsilon < \varepsilon_F$. At much higher temperatures (in the interior of stars, perhaps), there may be more electrons in excited states than there are below $\varepsilon_F$. We expect the *electronic heat capacity* $C_{el}$ to be larger at higher temperatures, that is, $C_{el} = f(T)$ should be a monotonically increasing function.

## 8.4 The Electronic Contribution to Solid-State Heat Capacity

In the section on classical wave equations, we found that the differential equation (Equation 2.4.7)

$$\frac{d^2 X(x)}{dx^2} + \beta^2 X(x) = 0 \tag{8.4.1}$$

leads to (among others) solutions of the form (section 2.5)

$$X(x) = B \sin \beta x. \tag{8.4.2}$$

Analogously, in the conventional particle in a box problem (McQuarrie 1983), the time-independent Schrödinger equation for a particle limited to motion along the $x$ direction,

$$\frac{d^2 \Psi(x)}{dx^2} + \frac{2mE}{\hbar^2} \Psi(x) = 0, \tag{8.4.3}$$

leads to the solution

$$\Psi(x) = B \sin \sqrt{\frac{2mE}{\hbar^2}} x. \tag{8.4.4}$$

Application of the boundary condition $\sin \beta x = 0$ at $x = L$ for wave motion constrained by barriers at $x = 0$ and $x = L$ leads to the equality

$$\beta L = n\pi$$

for the equation in $X(x)$ as seen in section 2.5. Due to the wave nature of subatomic particles (de Broglie), application of the same boundary condition to the solution of the Schrödinger equation for a particle executing translational motion along the $x$ axis between impenetrable barriers at $x = 0$ and $x = L$ takes the form $\sin \sqrt{2mE/\hbar^2}\, x = 0$ at $x = L$. This leads to

$$\sqrt{\frac{2mE}{\hbar^2}}L = n\pi. \tag{8.4.5}$$

Rearranging,

$$\frac{2mE}{\hbar^2}L^2 = n^2\pi^2, \tag{8.4.6}$$

or, bearing in mind that $\hbar = h/2\pi$,

$$E = \frac{n^2 h^2}{8mL^2}. \tag{8.4.7}$$

The generalization of this one-dimensional model to three dimensions (McQuarrie 1983) is straightforward, leading to

$$E = \frac{h^2}{8mL^2}\left(n_x^2 + n_y^2 + n_z^2\right), \tag{8.4.8}$$

where $n_x$, $n_y$, and $n_z$ are quantum numbers in each dimension of $x$, $y$, and $z$ space for a cubic restraining geometry.

## 8.5 The Ground State of a Fermi Gas

By the same reasoning we used in deriving the Rayleigh-Jeans equation, the number of orbitals (solutions to the Schrödinger equation) for particles confined to a cube approaches the volume of the positive octant of a sphere in *quantum number space*.

The radius vector in quantum number space is *n*. It has a magnitude

$$n = \sqrt{n_x^2 + n_y^2 + n_z^2} \tag{8.5.1}$$

by Pythagoras's theorem. The quantum numbers are restricted to positive values, and the number of orbitals is well approximated by the volume of one octant of a sphere

$$N_{orb} = \frac{1}{8}\frac{4}{3}\pi n^3 = \frac{\pi n^3}{6} \tag{8.5.2}$$

in quantum number space, provided that the number of orbitals is large. We wish to put electrons into all orbitals in the ground state of a Fermi gas $n_F$ confined to a cube of molar volume, so the number of electrons is

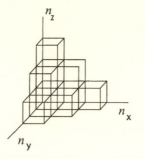

Figure 8.5.1. Orbitals in a three-dimensional quantum number space. Be careful to distinguish between Cartesian $x$, $y$, $z$ space and quantum number $n_x$, $n_y$, $n_z$ space.

twice the number of orbitals,

$$N_{el} = \frac{2\pi n_F^3}{6} = \frac{\pi n_F^3}{3}, \tag{8.5.3}$$

because each orbital accommodates two electrons (spin up, spin down).

From the general expression (equation 8.4.5) that we obtain by apply-ing boundary conditions to solutions of the Schrödinger equation in the particle in a box problem,

$$E = \frac{h^2}{8mL^2} n^2 = \frac{h^2}{2m} \left( \frac{n^2 \pi^2}{L^2} \right),$$

where $n$ is the magnitude of the radius vector in figure 8.5.1, we have

$$E = \frac{h^2}{2m} \left( \frac{\pi n}{L} \right)^2 \tag{8.5.4}$$

in general, or

$$\varepsilon_F = \frac{h^2}{2m} n_F^2 \left( \frac{\pi}{L} \right)^2 \tag{8.5.5}$$

for the special case of the energy of an electron on the surface of the Fermi sea, having a quantum number of $n_F$. From equation 8.5.3,

$$N_{el} = \frac{\pi n_F^3}{3}, \tag{8.5.6}$$

we have

$$n_F^2 = \left(\frac{3N_{el}}{\pi}\right)^{2/3} \qquad (8.5.7)$$

and, substituting this value of $n_F^2$ into the expression for $\varepsilon_F$,

$$\varepsilon_F = \frac{\hbar^2}{2m}\left(\frac{3N_{el}}{\pi}\right)^{2/3}\frac{\pi^2}{L^2} = \frac{\hbar^2}{2m}\left(\frac{3N_{el}}{\pi}\right)^{2/3}\left(\frac{\pi^3}{L^3}\right)^{2/3}$$

$$= \frac{\hbar^2}{2m}\left(\frac{3N_{el}\pi^2}{V}\right)^{2/3} \qquad (8.5.8)$$

where $V = L^3$ is the volume of the cubic restraining geometry.

## 8.6 The Number of Orbitals in the Ground State

Under the conditions specified, the discrete function $N_{orb} = f(n)$ can be treated as a continuous function

$$\frac{\delta N_{orb}}{\delta n} \cong \frac{dN_{orb}}{dn}$$

and we can use the result $N_{orb} = \pi n_F^3/6$, which is a special case for the filled ground state. In general, at any energy level, $N_{orb} = \pi n^3/6$, so

$$dN_{orb} = \frac{\pi n^2}{2}\,dn \qquad (8.6.1)$$

The number of orbitals in a small region of the quantum number space containing the vector $\mathbf{n}$ is shown in figure 8.6.1.

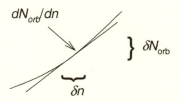

Figure 8.6.1. An infinitesimal increment in the number of orbitals with respect to the quantum number vector.

## 8.7 The Total Energy of Electrons in the Ground State

We would like to know the total energy of all of the electrons in the ground state of a Fermi gas. To find this, we multiply each orbital by $2\varepsilon_n$ where $\varepsilon_n$ is the energy of the orbital and there are two electrons per orbital. This gives the energy contribution from each orbital. The sum of energy contributions approaches an integral for large $n_F$. We then integrate each incremental energy from 0 to $n_F$, where $n_F$ is the upper limit of filled orbitals in the ground state:

$$U_0 = 2 \sum_0^{n_F} N_{\text{orb}} \varepsilon_n = \pi \int_0^{n_F} n^2 \varepsilon_n \, dn. \tag{8.7.1}$$

We have already solved for $\varepsilon_n$ (equation 8.5.4),

$$\varepsilon = \frac{\hbar^2}{2m} \left( \frac{\pi n}{L} \right)^2, \tag{8.7.2}$$

so the integral is

$$U_0 = \frac{\pi^3}{2m} \frac{\hbar^2}{L^2} \int_0^{n_F} n^4 \, dn = \frac{\pi^3}{10m} \left( \frac{\hbar}{L} \right)^2 n_F^5. \tag{8.7.3}$$

Because $N_{\text{el}} = \pi n_F^3 / 3$,

$$n_F^3 = \frac{3N_{\text{el}}}{\pi}, \tag{8.7.4}$$

the energy is

$$U_0 = \frac{\pi^3}{10m} \left( \frac{\hbar}{L} \right)^2 n_F^5 = \frac{\pi^3}{10m} \left( \frac{\hbar}{L} \right)^2 n_F^2 \left( \frac{3N_{\text{el}}}{\pi} \right)$$

$$= \frac{3\hbar^2}{10\,m} \left( \frac{\pi n_F}{L} \right)^2 N_{\text{el}} = \frac{3}{5} \left[ \frac{\hbar^2}{2m} \left( \frac{\pi n_F}{L} \right)^2 \right] N_{\text{el}}. \tag{8.7.5}$$

We recognize the term in square brackets as

$$\frac{\hbar^2}{2m} \left( \frac{\pi n_F}{L} \right)^2 = \varepsilon_F,$$

which leads to

$$U_0 = \frac{3}{5} N_{\text{el}} \varepsilon_F. \tag{8.7.6}$$

If we divide both sides of this equation by $N_{el}$, the result is the average energy per particle,

$$\overline{U}_0 = \frac{3}{5}\varepsilon_F. \tag{8.7.7}$$

Because $\varepsilon_F$ is inversely related to $L^2$, the energy increases as $L$ decreases, that is, Fermi forces are repulsive. Bear in mind that this is generally true of fermions and has nothing to do with charge repulsion.

At this point let us digress for a moment to observe that

$$n_F^2 = \left(\frac{3N_{el}}{\pi}\right)^{2/3}$$

gives

$$n_F = \left(\frac{3N_A}{2\pi}\right)^{1/3} \cong 6.6 \times 10^7$$

as the number of states occupied by spin-up and spin-down electrons. Thus, the condition of many orbitals in figure 8.5.1 is amply satisfied.

## 8.8 The Density of States

The density of quantum states $\rho_{states}(\varepsilon) = dN_{orb}/d\varepsilon$ per unit energy increment between $\varepsilon$ and and $\varepsilon + d\varepsilon$ is, from equation 8.6.1,

$$\rho_{states}(\varepsilon)\,d\varepsilon = \frac{\pi n^2}{2}\frac{dn}{d\varepsilon}\,d\varepsilon. \tag{8.8.1}$$

Because $\varepsilon$ goes up as $n^2$, the plot of the number of quantum states, $n$ versus $\varepsilon$, is a parabolic arc in the positive quadrant of figure 8.8.1.

From equation 8.7.2

$$\varepsilon = \frac{\hbar^2}{2m}\left(\frac{\pi n}{L}\right)^2 \tag{8.8.2}$$

we solve for $n$,

$$n = \left(\frac{2mL^2\varepsilon}{\pi^2\hbar^2}\right)^{1/2}, \tag{8.8.3}$$

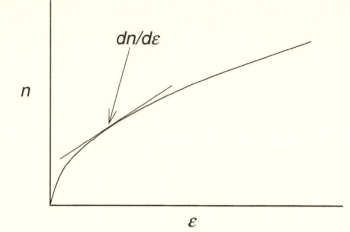

Figure 8.8.1. The locus of states is a parabolic arc $n$ versus $\varepsilon$.

and find

$$\frac{dn}{d\varepsilon} = \frac{1}{2}\left(\frac{2mL^2\varepsilon}{\pi^2\hbar^2}\right)^{-1/2}\left(\frac{2mL^2}{\pi^2\hbar^2}\right) = \left(\frac{mL^2}{2\pi^2\hbar^2\varepsilon}\right)^{1/2}. \tag{8.8.4}$$

This gives

$$n^2\frac{dn}{d\varepsilon} = \left(\frac{2mL^2\varepsilon}{\pi^2\hbar^2}\right)\left(\frac{mL^2}{2\pi^2\hbar^2\varepsilon}\right)^{1/2}$$

$$= \frac{2m^{3/2}L^3\varepsilon}{\sqrt{2}\pi^3\hbar^3\sqrt{\varepsilon}} = \frac{\sqrt{2}m^{3/2}L^3}{\pi^3\hbar^3}\sqrt{\varepsilon}. \tag{8.8.5}$$

If we multiply by $2\pi$, divide by 2 (from equation 8.8.1), and note that $L^3 = V$, where $V$ is the volume of the cube in quantum number space, the density of states is

$$\rho_{\text{states}}(\varepsilon) = \frac{V}{2\pi^2}\left(\frac{2m}{\hbar^2}\right)^{3/2}\sqrt{\varepsilon} \tag{8.8.6}$$

for electrons.

The density of *occupied* orbitals is $\bar{n}_i\rho_{\text{states}}(\varepsilon)$.

At 0 K in figure 8.8.2, the density of filled states is given by the area under the parabola from zero to the vertical cutoff $\varepsilon_F$. At $T = 0.20T_F$, an inverse sigmoidal distribution replaces the vertical cutoff at $\varepsilon_F$. In figure 8.8.2, the inverse sigmoidal curve of figure 8.3.1 superimposed on the

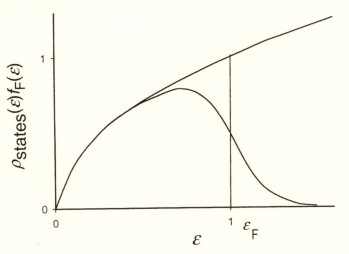

Figure 8.8.2. The density of filled energy states at $T = 0\,\text{K}$ and $T = 0.20T_F$.

parabolic arc of figure 8.8.1, that is, the continuous curve in figure 8.8.2 is the product $\bar{n}_i\rho_{\text{states}}(\varepsilon) = f_F(\varepsilon)\rho_{\text{states}}(\varepsilon)$, where $f_F(\varepsilon)$ is the Fermi-Dirac distribution function in the summary equation 7.20.4.

The total number of electrons in the system is

$$N = \int_0^\infty \rho_{\text{states}}(\varepsilon)f_F(\varepsilon)\,d\varepsilon \tag{8.8.7}$$

and the total energy is

$$U = \int_0^\infty \varepsilon\rho_{\text{states}}(\varepsilon)f_F(\varepsilon)\,d\varepsilon. \tag{8.8.8}$$

In the special case of the ground state,

$$N = \int_0^{\varepsilon_F} \rho_{\text{states}}(\varepsilon)f_F(\varepsilon)\,d\varepsilon \tag{8.8.9}$$

and

$$U = \int_0^{\varepsilon_F} \varepsilon\rho_{\text{states}}(\varepsilon)f_F(\varepsilon)\,d\varepsilon, \tag{8.8.10}$$

where the upper limit of integration $\varepsilon_F$ is the energy at the Fermi level.

## 8.9 The Energy of an Electron Gas

Notice that, in the general expression for the energy, $U = \int_0^\infty \varepsilon \rho_{\text{states}}(\varepsilon)$ $f_F(\varepsilon)\,d\varepsilon$, the Fermi-Dirac distribution function $f_F(\varepsilon)$ is unique to each temperature. Even though the limits of integration do not change, the integral yields a different value for each $f_F(\varepsilon)$ at each temperature. This coincides with our expectation that $U$ is a function of $T$. Knowing the energy of a Fermi gas at any temperature, $\int_0^\infty \varepsilon \rho_{\text{states}} f_F(\varepsilon)\,d\varepsilon$, and the energy of the ground state, we can find the energy change on heating the Fermi gas from 0 to $T$ K,

$$\Delta U = \int_0^\infty \varepsilon \rho_{\text{states}}(\varepsilon) f_F(\varepsilon)\,d\varepsilon - \int_0^{\varepsilon_F} \varepsilon \rho_{\text{states}}(\varepsilon) f_F(\varepsilon)\,d\varepsilon \qquad (8.9.1)$$

We have an expression for $N$,

$$N = \int_0^\infty \rho_{\text{states}}(\varepsilon) f_F(\varepsilon)\,d\varepsilon, \qquad (8.9.2)$$

and we know that the number of electrons does not change during the heating process, so

$$N = \int_0^\infty \rho_{\text{states}}(\varepsilon) f_F(\varepsilon)\,d\varepsilon = \int_0^{\varepsilon_{FD}} \rho_{\text{states}}(\varepsilon) f_F(\varepsilon)\,d\varepsilon. \qquad (8.9.3)$$

Heat absorbed by an electron gas can be thought of as energy contributing to the sum of two-electron promotion processes. First, an electron *near* the surface of the Fermi sea must be promoted *to* its surface. This requires an energy that is the difference between the final energy of the electron and its initial energy, $(\varepsilon_F - \varepsilon)$. To find the energy input for many electrons at varying depths in the electron sea, we integrate over the energy for one electron times the density of states of the electrons *that are being removed* from the Fermi sea, $\rho_{\text{states}}(1 - f_F(\varepsilon))$. This integral corresponds to the area $C = A - B$ in figure 8.9.1 up to the limit $\varepsilon_F$. The integral is

$$\int_0^{\varepsilon_F} (\varepsilon_F - \varepsilon)\,\rho_{\text{states}}(\varepsilon) \left(1 - f_F(\varepsilon)\right) d\varepsilon, \qquad (8.9.4)$$

where the limits are over all electrons with energies from 0 to $\varepsilon_F$.

Once at the surface of the Fermi sea, some electrons are promoted to higher-energy states. This time the integral corresponds to the energy $\varepsilon - \varepsilon_F$ times the density of states *above* the surface of the sea, which is

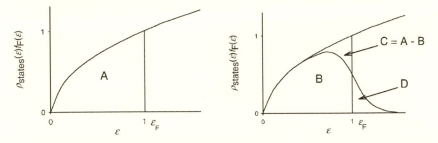

Figure 8.9.1. Electron energy promotion from the Fermi sea to an excited state.

area D in figure 8.9.1, between $\varepsilon_F$ and $\infty$. The integral is

$$\int_{\varepsilon_F}^{\infty} (\varepsilon - \varepsilon_F)\, \rho_{\text{states}}(\varepsilon) f_F(\varepsilon)\, d\varepsilon. \tag{8.9.5}$$

The sum of energies for this hypothetical two-step process is

$$\Delta U = \int_0^{\varepsilon_F} (\varepsilon_F - \varepsilon)\left(1 - f_F(\varepsilon)\right) \rho_{\text{states}}(\varepsilon)\, d\varepsilon$$

$$+ \int_{\varepsilon_F}^{\infty} (\varepsilon - \varepsilon_F)\, \rho_{\text{states}}(\varepsilon) f_F(\varepsilon)\, d\varepsilon. \tag{8.9.6}$$

## 8.10 The Low-Temperature Heat Capacity of an Electron Gas

From the expression for $\Delta U$, equation 8.9.6,

$$C_{\text{el}} = \frac{d}{dT}(U - U_0) = \frac{dU}{dT} = \int_0^{\infty} (\varepsilon - \varepsilon_F)\, \rho_{\text{states}}(\varepsilon) \frac{df_F(\varepsilon)}{dT}\, d\varepsilon \tag{8.10.1}$$

because $f_F(\varepsilon)$ is the only thing in the equation that depends on $T$. The Fermi temperature is very high. At normal ambient temperatures, $k_B T/\varepsilon_F \ll 0.01$. The first derivative of $f_F(\varepsilon)$ with respect to $T$ is large (and negative) only near $\varepsilon_F$, which is the vertical inflection point in the curve in figures 8.3.1 and 8.8.2. Near $\varepsilon_F$, the integral for the heat capacity becomes

$$C_{\text{el}} = \rho_{\text{states}}(\varepsilon_F) \int_0^{\infty} (\varepsilon - \varepsilon_F) \frac{df_F(\varepsilon)}{dT}\, d\varepsilon. \tag{8.10.2}$$

From the Fermi distribution, equation 7.20.4,

$$f_F(\varepsilon) = \left(e^{(\varepsilon_i - \varepsilon_F)/k_B T} + 1\right)^{-1}, \tag{8.10.3}$$

we find

$$\frac{df_F(\varepsilon)}{dT} = \frac{(\varepsilon - \varepsilon_F)}{k_B T^2} \frac{e^{(\varepsilon_i - \varepsilon_F)/k_B T}}{\left(e^{(\varepsilon_i - \varepsilon_F)/k_B T} + 1\right)^2}. \qquad (8.10.4)$$

To evaluate the integral and find the electronic heat capacity, we make the simplifying substitution $x = (\varepsilon - \varepsilon_F)/k_B T$ and drop the subscript $i$, which converts

$$C_{el} = \rho_{states}(\varepsilon_F) \int_0^\infty (\varepsilon - \varepsilon_F) \frac{(\varepsilon - \varepsilon_F)}{k_B T^2} \frac{e^{(\varepsilon_i - \varepsilon_F)/k_B T}}{\left(e^{(\varepsilon_i - \varepsilon_F)/k_B T} + 1\right)^2} d\varepsilon \qquad (8.10.5)$$

into

$$C_{el} = \rho_{states}(\varepsilon_F) \int_0^\infty x \frac{k_B^2 T^2 x}{k_B T^2} \frac{e^x}{(e^x + 1)^2} d\varepsilon \qquad (8.10.6)$$

under the approximation that $\varepsilon$ is not too far from $\varepsilon_F$. This leads to

$$C_{el} = \rho_{states}(\varepsilon_F) k_B^2 T \int_{-\infty}^\infty x^2 \frac{e^x}{(e^x + 1)^2} dx \qquad (8.10.7)$$

because $d\varepsilon = k_B T \, dx$. The integral $\int_{-\infty}^\infty x^2 [e^x/(e^x + 1)^2] \, dx = \pi^2/3$ is not easy to find in the tables. It can be verified by numerical integration (be careful of the point $x = 0$). Including the numerical value of the integral, we get

$$C_{el} = \rho_{states}(\varepsilon_F) k_B^2 T \frac{\pi^2}{3} = 3.29 \rho_{states}(\varepsilon_F) k_B^2 T \qquad (8.10.8)$$

which is linear in $T$.

## 8.11 The Debye-Sommerfeld Equation

Having found that the electronic contribution to the heat capacity is linear in $T$, we can add the electronic contribution to the Debye equation, to obtain the heat capacity as the sum of *lattice* (Debye) vibrations in a metal plus an electronic contribution:

$$C_v = 9Lk_B \left(\frac{T}{\Theta}\right)^3 \int_0^{x_m} \frac{x^4 e^x}{(e^x - 1)^2} dx + \rho_{states}(\varepsilon_F) k_B^2 \frac{\pi}{3} T. \qquad (8.11.1)$$

By arguments given in the section on the Debye equation, the sum in equation 8.11.1 becomes

$$C_v = 1944 \left( \frac{T}{\Theta_D} \right)^3 + \gamma T \qquad (8.11.2)$$

where $\gamma$ is the proportionality constant $\rho_{states}(\varepsilon_F) k_B^2 \pi / 3$. Dividing by $T$ we get

$$\frac{C_v}{T} = \left( \frac{1944}{\Theta_D^3} \right) T^2 + \gamma, \qquad (8.11.3)$$

which leads to both $\Theta_D$ and $\gamma$ obtained from the slope and intercept of an experimental plot of $C_v/T$ versus $T^2$ at very low temperatures.

In figure 8.11.1 a linear least-squares treatment of the experimental data for silver yields an intercept of $6.199 \times 10^{-4} \, J \, mol^{-1} \, K^{-2}$ and a slope of $1.561 \times 10^{-4} \, J \, mol^{-1} \, K^{-4}$. This leads to $\gamma = 6.2 \times 10^{-4} \, J \, mol^{-1} \, K^{-2}$ and

$$\frac{1944}{\Theta_D^3} = 1.561 \times 10^{-4} \, J \, mol^{-1} \, K^{-4},$$

$$\Theta_D^3 = \frac{1944}{1.561 \times 10^{-4}} = 1.245 \times 10^7,$$

$$\Theta_D = 232 \, K$$

from the slope. The temperature $\Theta_D = 232 \, K$ compares with $\Theta_D(\text{silver}) = 225 \, K$ calculated from empirical curve fits to the Debye equation at higher temperatures, without taking into account the electronic contribution to $C_V$.

The intercept $\gamma$,

$$C_{el} = \gamma T = 3.29 \rho(\varepsilon_F)_{states} k_B^2 T,$$

can be calculated from the density of states at the Fermi level,

$$\rho(\varepsilon_F)_{states} = \frac{3N}{2\varepsilon_F} = \frac{3N}{2k_B T_F},$$

where $T_F$ is an experimental value, $6.4 \times 10^4 \, K$ for silver. This alternative method gives

$$\frac{C_{el}}{T} = 3.29 \left( \frac{3}{2} \right) \frac{N_A k_B}{(6.4 \times 10^4)} = 6.41 \times 10^{-4} \, J \, K^{-2} \, mol^{-1}$$

as compared to the experimental value from figure 8.11.1 of $6.2 \times 10^{-4} \, J \, K^{-2} \, mol^{-1}$.

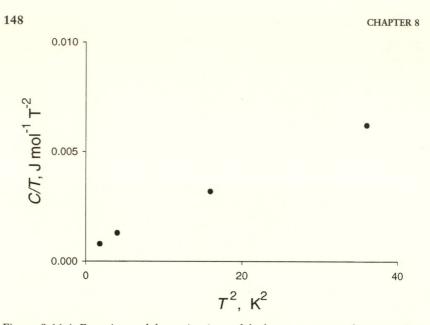

Figure 8.11.1. Experimental determinations of the low-temperature heat capacity divided by $T$ versus $T^2$ for silver.

## PROBLEMS

**8.1.** What is the number density of electrons in a molar volume of lithium, Li? Give your answer in both electrons per $m^3$ and electrons per $cm^3$. The density of Li is $0.534\,\mathrm{g\,cm^{-3}}$.

**8.2.** What is the Fermi temperature of a molar volume of Li?

**8.3.** Verify the equality

$$C(C-1)\ \cdots\ (C-N+1) = \frac{C!}{(C-N)!}$$

for three simple examples.

**8.4.** The number of ways that $N$ distinguishable objects can be placed in $C$ numbered cells with no more than one object per cell is

$$W = \frac{C!}{(C-N)!}$$

(a) Give a logical argument as to why this equation is true.

(b) Find $W$ for the same conditions except that the objects are indistinguishable. Give a logical argument as to why your equation is true.

**8.5.** What is the (Fermi) electron pressure within a molar volume of Li?

# Nine

## Consequences of the Bose-Einstein Distribution

THE EARLIEST AND, in some ways, still the most remarkable application of the Bose-Einstein statistic goes all the way back to Bose's derivation of Planck's blackbody radiation equation. Prior derivations, including Einstein's 1907 paper, mixed a part $(8\pi v^2/c^3)\,dv$ derived classically by Rayleigh and Jeans with a quantized energy $E$ to arrive at $\rho_v\,dv$ for the density of radiation within a blackbody cavity. It was twenty four years after Planck's original derivation that Bose showed how Planck's equation can be obtained in an internally consistent, wholly quantum mechanical derivation. In the end, getting the right answer was a matter of correctly subdividing quantum number space and learning how to count indistinguishable objects to find the configuration of maximum probability.

### 9.1 Of Waves and Particles

Which model of blackbody radiation is correct? Is thermal radiation trapped inside a blackbody cavity a sinusoidal oscillation in the electromagnetic field (electromagnetic radiation) constrained to certain wavelengths to avoid destructive interference, or is it a bunch of harmonic oscillators excited to one or another energy level, or is it many energy levels each populated by some number of particles called photons? Richard Feynman put the situation most clearly:

> From one point of view, we can analyze the electromagnetic field in a box or cavity in terms of a lot of harmonic oscillators, treating each mode of oscillation according to quantum mechanics as a harmonic oscillator. From a different point of view, we can analyze the same physics in terms of identical Bose particles. And the results of both ways of working *are always in exact agreement*. There is no way to make up your mind whether the electromagnetic field is really to be described as a quantized harmonic oscillator or by giving how many photons there are in each condition. The two views turn out to be mathematically identical. (Feynman, et al. 1965, vol. III, pp. 4–9)

## 9.2 Bose: The Density of Photon Modes

By analogy with our calculation of the density of states in chapter 8, the equation for the energy spectrum of a particle in a field-free cubic box, equation 8.5.4,

$$\varepsilon = \frac{n^2 h^2}{8mL^2} = \frac{h^2}{2m}\left(\frac{\pi n}{L}\right)^2 \tag{9.2.1}$$

gives

$$\frac{2m\varepsilon}{h^2} = \frac{\pi^2}{L^2}n^2. \tag{9.2.2}$$

From equation 2.10.2 c, a general expression relating frequency and wave propagation speed v is

$$\frac{\omega}{v} = \frac{\pi}{L}n, \tag{9.2.3}$$

where $\omega$ is the frequency of a standing wave in a quantum number space spanned by the components of radius vector $n$ of magnitude $n$. For a small increment in $\omega$

$$\frac{d\omega}{c} = \frac{\pi}{L}dn \tag{9.2.4}$$

for photons of speed $c$.

The volume in quantum number space goes up as the volume of one octant of a sphere of radius $n$ (section 4.3)

$$\frac{1}{8}4\pi n^2 dn. \tag{9.2.5}$$

If one quantum state is, according to Bose, a "cell of magnitude $h^3$" in quantum number space, the number of quantum states $N_\omega$ goes up as

$$\frac{1}{8}4\pi\left(\frac{\omega L}{c\pi}\right)^2\frac{Ld\omega}{c\pi} = \frac{V\omega^2}{2\pi^2 c^3}d\omega \tag{9.2.6}$$

or

$$dN_\omega = \frac{V\omega^2}{2\pi^2 c^3}d\omega. \tag{9.2.7}$$

After multiplying by 2 for the polarization of photons, we have a purely quantum mechanical expression for $\rho_{\text{photon states}}$, the number of modes in unit frequency range,

$$\rho_{\text{photon states}} = \frac{dN_\omega}{d\omega} = \frac{V}{\pi^2 c^3}\omega^2 \qquad (9.2.8)$$

where $V = L^3$ is the volume in quantum number space. Bose, who worked in units of $v$ rather than angular frequency $\omega = 2\pi v$, wrote this as

$$A^s = \frac{8\pi v^{s2}}{c^3}dv^s \qquad (9.2.9)$$

and described $A^s$ as "those cells which belong to the frequency interval $dv^s$."

Using a combinatorial argument (see the problems at the end of this chapter) Bose separated cells into groups containing no photons $p_0^s$, one photon $p_1^s$, two photons $p_2^s$, and so on, and showed that the number of distributions is $A^s/\prod p_r^s$ from which the probability of a state defined by all the $p_r^s$ is

$$\prod_s \frac{A^s}{\prod p_r^s}. \qquad (9.2.10)$$

This might seem a discouragingly small probability for the enormous number of possibilities, but one distribution so dominates all the others that only it needs to be taken into account. (Notice that this argument works only for bosons because fermions do not double up.)

Bose then used Lagrange's method of undetermined multipliers to arrive at the most probable distribution of the number of photons in an interval $dv^s$:

$$N^s = \sum_r rp_r^s = \sum_r rA^s\left(1 - e^{-hv^s/\beta}\right)e^{-hv^s/\beta} = \frac{A^s e^{-hv^s/\beta}}{1 - e^{-hv^s/\beta}}, \qquad (9.2.11)$$

which gives the energy

$$E = \sum_s \frac{8\pi hv^{s3}\,dv^s}{c^3}V\frac{e^{-hv^s/\beta}}{1 - e^{-hv^s/\beta}} \qquad (9.2.12)$$

where $8\pi hv^{s3}\,dv^s/c^3$ is his expression for photon states. After evaluating $\beta = kT$ by a short thermodynamic argument (Bose took $1/\beta$ as one of his undetermined multipliers in the Lagrangian procedure), he got

$$E = \sum_s \frac{8\pi hv^{s3}}{c^3}V\frac{1}{e^{-hv^s/kT} - 1}\,dv^s \qquad (9.2.13)$$

leading to Planck's equation for the energy density $E/V = u(v) = \bar{n}hv\rho(v)_{\text{photon states}}$, which we wrote as equation 4.5.25, the result of the Planck-Einstein derivation,

$$u_v(v, T) = \frac{8\pi v^2}{c^3} \frac{hv}{e^{hv/k_B T} - 1}. \tag{9.2.14}$$

The number distribution of bosons at a fixed temperature over an energy spectrum of equidistant levels is

$$\bar{n}_i = \frac{1}{e^{(\varepsilon_i - \mu)/k_B T} - 1} \tag{9.2.15}$$

in general or

$$\bar{n}_i = \frac{1}{e^{hv/k_B T} - 1} \tag{9.2.16}$$

for the specific case of the photon distribution.

The known laws of radiation follow from the Bose derivation (as they must). For example, the total electromagnetic energy within a blackbody cavity is

$$U(T) = \frac{\pi^2 k_B}{15\hbar^3 c^3} T^4 = 7.564 \times 10^{-16} T^4 \, \text{J} \, \text{m}^{-3}$$

and by the geometric argument previously given (section 3.5),

$$\sigma = 5.669 \times 10^{-8} \, \text{J} \, \text{s}^{-1} \, \text{m}^{-2} \, \text{K}^{-4}, \tag{9.2.17}$$

which is the Stefan-Boltzmann constant.

## 9.3 Why Is $\mu = 0$ for Photons?

In going from the equation for $\bar{n}_i = 1/(e^{(\varepsilon_i - \mu)/k_B T} - 1)$ (equation 9.2.15) to $\bar{n}_i = 1/(e^{hv/k_B T} - 1)$ (equation 9.2.16), it is evident that we have set $\mu$, the chemical potential, equal to zero. Why is this so? The short answer is that $\mu = 0$ for radiation in equilibrium with its surroundings because photons are not conserved. Let us see why in more detail. Consider an isothermal reversible expansion carried out by an ideal photon gas in a cylinder thermostated by an infinite thermal bath at temperature $T$ (figure 9.3.1).

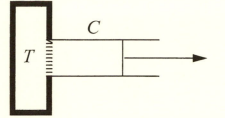

Figure 9.3.1. Cylinder $C$ containing a photon gas in contact with a thermal reservoir at temperature $T$. Heat but not matter can be transferred across the dashed line.

From the first law of thermodynamics, the energy of a system varies (section 7.26) according to

$$dU = dq - p\,dV \qquad (9.3.1)$$

for an infinitesimal expansion or contraction $dV$ where only pressure-volume work is done by the system. The quantity $dq$ is the amount of heat exchanged with the surroundings. From the second law of thermodynamics, the entropy $S$ for a reversible heat transfer is

$$dS = \frac{dq_{rev}}{T} \qquad (9.3.2)$$

so we arrive at the fundamental equation for $dU$ of a one-component system

$$dU = TdS_{rev} - pdV, \qquad (9.3.3)$$

where $p$ is the pressure against which the system expands or contracts.

Heat flow from the reservoir to the photon gas within the cylinder compensates for the energy lost in an expansion, but the only way this can happen is for the reservoir to supply more photons to the photon gas within the cylinder. The reverse is true for a contraction. The cylinder contains only photons at the beginning of the process and only photons at the end but their number is not the same. Unlike atoms, photons are not conserved.

From the Gibbs generalization of the fundamental equation for $dU$,

$$dU = T\,dS_{rev} - p\,dV + \mu\,dN, \qquad (9.3.4)$$

where $\mu$ is the chemical potential, we see that these two equations are equal only if

$$\mu\,dN = 0, \qquad (9.3.5)$$

but we know that $dN \neq 0$ because photons are being created or destroyed; therefore,

$$\mu = 0. \tag{9.3.6}$$

## 9.4 Phonons

By Feynman's equivalence principle (section 9.1), sound waves propagating through an elastic solid can be treated as though they are particles distributed over a spectrum of harmonic oscillator energy levels. The new particles are called *phonons* because they transmit sound energy.

The density of phonons is

$$\rho_{\text{phonon}} = \frac{3}{2} \frac{V}{\pi^2 v_s^3} \omega^2 \tag{9.4.1}$$

where the phonon density is a function of frequency, $\rho(\omega)$. The expression for phonon density is the same as that for photon density (equation 9.2.8) except for the factor 3/2 included to account for the fact that phonons traveling at speed $v_s$ have two transverse and one longitudinal amplitude, in contrast to electromagnetic waves, which have only transverse amplitudes.

The number of modes of vibration, $3N$, where $N$ is the number of atoms in a solid sample, leads to

$$3N = \int_0^{\omega_D} \rho(\omega)_{\text{phonon}} \, d\omega = \frac{V}{2\pi^2 v_s^3} \omega_D^3 \tag{9.4.2}$$

where $\omega_D$ is the Debye cutoff frequency. This gives

$$\omega_D = \left( \frac{6\pi^2 N v_s^3}{V} \right)^{1/3} \tag{9.4.3}$$

or

$$\frac{6\pi^2 N v_s^3}{\omega_D^3} = V, \tag{9.4.4}$$

but $\omega/V$ is the density of states for phonons; hence, substituting

$$\rho(\omega)_{\text{phonon}} = \frac{3}{2\pi^2 v_s^3} \omega^2 \left( \frac{6\pi^2 N v_s^3}{\omega_D^3} \right)$$

$$= 9N \left( \frac{\omega^2}{\omega_D^3} \right). \tag{9.4.5}$$

The total energy of a solid at temperature $T$ is

$$U(T) = \int_0^{\omega_D} \rho(\omega)_{\text{phonon}} \bar{n} \hbar \omega \, d\omega$$

$$= \frac{9N\hbar}{\omega_D^3} \int_0^{\omega_D} \frac{\omega^2 d\omega}{e^{\hbar\omega/k_B T} - 1} \qquad (9.4.6)$$

where $\bar{n}$ is given by the Planck distribution. This integration has been done before:

$$U(T) = \frac{9NT^4}{(\hbar\omega_D)^3} \int_0^x \frac{x^3 dx}{e^x - 1} = \frac{9NT^4}{(\hbar\omega_D)^3} \left(\frac{\pi^4}{15}\right)$$

$$= \frac{3\pi^4 Nk_B}{5\Theta_D} T^4 \qquad (9.4.7)$$

with a molar heat capacity

$$C_V = \left(\frac{\partial U}{\partial T}\right)_V = \frac{12\pi^4 N_A k_B}{5} \left(\frac{T}{\Theta_D}\right)^3 = 1944 \left(\frac{T}{\Theta_D}\right)^3 \qquad (9.4.8)$$

at low temperature, in agreement with equation 6.5.3.

## 9.5  The Influence of Symmetry Numbers on Rotational Spectroscopy

When electromagnetic radiation of frequency $\nu_{\text{rad}}$ falls on matter, molecules may or may not be promoted from a lower state to a higher one, according to whether the energy of the radiation $\varepsilon_{\text{rad}} = h\nu_{\text{rad}}$ matches an energy gap $\Delta\varepsilon$ between levels within the molecules which we (properly) regard as mechanical systems. Energy spacings are governed by quantum mechanics. For example, in the case of a harmonic oscillator, vibrational energies are found at

$$\varepsilon_{\text{vib}} = \left(n_{\text{vib}} + \frac{1}{2}\right) h\nu_{\text{vib}}, \qquad n_{\text{vib}} = 1, 2, 3, \ldots, \infty, \qquad (9.5.1)$$

where $n_{\text{vib}}$ is a vibrational quantum number. The vibrational frequency of a diatomic molecule is governed by the force constant $k$ of its bond and the reduced mass $\mu$ of its atoms through the equation $\nu_{\text{vib}} = (1/2\pi)\sqrt{k/\mu}$ from chapter 2. The natural unit for $\nu$ is Hz $= s^{-1}$ (i.e., cycles per second) but spectroscopists frequently report frequencies in $\bar{\nu}\,\text{cm}^{-1}$ where $\bar{\nu} = \nu/c$. For our purposes it is convenient to multiply by $10^{-2}$ to obtain $\bar{\nu}$ in units of reciprocal meters, $\text{m}^{-1}$.

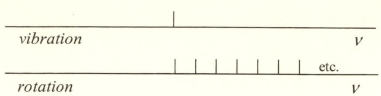

Figure 9.5.1. Schematic diagrams of vibrational and rotational spectra.

The energy gap between *adjacent* vibrational energy levels is $\Delta\varepsilon_{vib} = h\nu_{vib}$ from equation 9.5.1. For absorption of radiant energy, if the lower energy level is occupied and if certain selection rules are satisfied, molecules may be driven from the lower vibrational state to the adjacent higher state with $\varepsilon_{rad} = \Delta\varepsilon_{vib} = h\nu_{rad}$.

The pure vibrational spectrum of a harmonic oscillator is very simple, consisting of only one line at the resonance frequency $\nu_{rad} = \nu_{vib}$. This is because the spacings between energy levels in the harmonic oscillator approximation are all equal. At the resonance frequency, molecules are promoted by one vibrational quantum level to the next higher level, no matter where they start out.

We can also measure the spacing between rotational energy levels $\Delta\varepsilon_{rot}$ by finding $\nu_{rot}$ from the resonance frequency $\nu_{rad} = \nu_{rot}$ of the *lowest*-energy line in a pure rotational absorption spectrum,

$$h\,\nu_{rot} = \frac{h^2}{8\pi^2 I}(J'(J'+1) - J(J+1)) = 2\frac{h^2}{8\pi^2 I}. \qquad (9.5.2)$$

The quantum numbers $J' = 1$ and $J = 0$ for this transition lead to

$$\nu_{rot} = \frac{h}{4\pi^2 I}. \qquad (9.5.3)$$

Here $I$ is the moment of inertia, $I = \mu r^2$, $\mu$ is the reduced mass, $1/\mu = 1/m_1 + 1/m_2$, and $m$ designates the atomic mass of atom 1 and atom 2 separated by bond length $r$ in a diatomic molecule. Knowing $\nu_{rot}$ for the lowest line, we can calculate the entire manifold of rotational energy levels and hence the entire (idealized) rotational absorption spectrum by substituting different integer values into equation 9.5.2. Energy, and consequently frequency, increases from left to right in figure 9.5.1. Following up the sequence $J(J + 1)$ starting at 0, we get $0, 2, 6, 12, 20, \ldots$ for the energy levels. This gives differences of $2, 4, 6, 8$, and so on, and leads to the even spacing in the diagram. Each transition from $J$ to $J + 1$ involves an energy spacing that is $2h/8\pi^2 I$ higher than the one below it and gives a line at a frequency $2h/8\pi^2 I$ to the right of the one below it.

There are several factors ignored in this idealized treatment of vibrational and rotational spectra which cause the real spectra to be considerably more complicated than we have indicated here. One of these is the *symmetry number* (see below). There are other mechanical factors as well, such as centrifugal stretching of rotating molecules and anharmonic vibrations, but we will not devote any attention to them because they do not alter elementary spectroscopic principles.

## 9.6 The Vibrational Partition Function $q_{vib}$

In treating vibrational motion of a diatomic molecule assumed to be a harmonic oscillator, the partition function $q_{vib}$ has been defined as a sum (chapter 7)

$$q_{vib} = \sum_{levels} e^{-\beta \varepsilon_{vib}} = 1 + e^{-\beta \varepsilon_{vib}} + e^{-2\beta \varepsilon_{vib}} + e^{-3\beta \varepsilon_{vib}} + \cdots , \qquad (9.6.1)$$

whence

$$q_{vib} = \frac{1}{1 - e^{-\beta \varepsilon_{vib}}} \qquad (9.6.2)$$

where $\varepsilon_{vib} = h\nu$. For notational convenience, we also define a *vibrational characteristic temperature*

$$\theta_{vib} = \frac{\varepsilon_{vib}}{k_B} = \frac{h\nu}{k_B} \qquad (9.6.3)$$

In terms of $\theta_{vib}$,

$$q_{vib} = \frac{1}{1 - e^{-\beta h \nu_{vib}}} = \frac{1}{1 - e^{-\theta_{vib}/T}} . \qquad (9.6.4)$$

The partition function and the characteristic temperature tell us something about the number distribution of molecules over energy levels or states. If the temperature of a system is not too far from $\theta$, the higher states are appreciably populated. If not, they are largely vacant.

The vibrational characteristic temperature of many simple diatomic molecules, $H_2$, $N_2$, CO, and so on, is far above ambient temperature of 298 K. For these examples, $\theta_{vib} = 5987$ K, $3352$ K, and $3084$ K, respectively. This means that their higher vibrational levels are not appreciably populated and

$$q_{vib} = \sum e^{-\theta_{vib}/T} = 1 + e^{-2big} + e^{-2big} + e^{-3big} + \cdots \qquad (9.6.5)$$

where the $e^{-\text{big}}$ terms are $e$ taken to a large negative exponent. The terms beyond the first term are essentially zero. They drop out and, for quantized vibrational motion,

$$q_{\text{vib}} = 1.0. \qquad (9.6.6)$$

## 9.7 The Rotational Partition Function $q_{\text{rot}}$

Rotational energy levels are complicated by degeneracy. There may be more than one quantum state at the same energy level. A diatomic molecule rotates about an axis perpendicular to its plane of rotation just as a bicycle wheel rotates about its axis perpendicular to the plane of the wheel. The rotational partition function is

$$q_{\text{rot}} = \sum_{\text{levels}} (2J + 1)e^{-\beta(J(J+1)h^2/8\pi^2 I)}. \qquad (9.7.1)$$

By quantum mechanical arguments, one finds that there are $2J + 1$ spatial *orientations* that the axis of rotation can take. For example, when $J = 1, 2J + 1$ gives three orientations (figure 9.7.1).

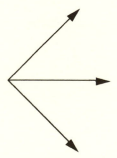

Figure 9.7.1. Three rotational orientations for $J = 1$.

In the absence of an external field, these three orientations have the same energy. There are three *states* at the $J = 1$ *level*. The level has a threefold degeneracy. Obtaining the partition function by summing over levels requires that the number of states at each level be taken into account. This is the origin of the $(2J + 1)$ term in equation 9.7.1.

Notation is becoming tedious at this point so we define a *rotational characteristic temperature* $\theta_{\text{rot}}$

$$\theta_{\text{rot}} = \frac{h^2}{8\pi^2 I k_B}, \qquad (9.7.2)$$

whereupon

$$q_{rot} = \sum_{levels} (2J + 1)e^{-J(J+1)\theta_{rot}/T} \qquad (9.7.3)$$

where $T$ is the temperature in kelvins. We usually make the approximation that the rotational levels are so close together that they can be treated as a continuum. This permits the summation (9.7.3) to be replaced by an integral over $J$

$$q_{rot} = \int_0^\infty (2J + 1)e^{-J(J+1)\theta_{rot}/T} \, dJ. \qquad (9.7.4)$$

The standard integral $\int_0^\infty e^{-au} \, du = 1/a$ where $u = J(J + 1)$ gives

$$q_{rot} = \frac{T}{\theta_{rot}}. \qquad (9.7.5)$$

Rotational characteristic temperatures are considerably lower than vibrational characteristic temperatures for diatomic molecules, indicating that diatomic molecules are distributed over a wider range of rotational levels than over vibrational levels under ordinary laboratory conditions. In other words, rotational levels are closer together. Some characteristic temperatures for rotation are $H_2$, 175, HCl, 30.5, $O_2$, 4.9, and $I_2$, 0.11 K.

In making the continuum approximation for rotational motion, we are pushing things a bit. For most diatomic molecules at most temperatures, the approximation is valid, but it is not very good at low temperatures, especially for light molecules like hydrogen. At low temperatures, the symmetry number has a profound effect on $q_{rot}$.

## 9.8 Symmetry Numbers

Classically, one can understand what a symmetry number is by considering a homonuclear diatomic molecule, like molecular hydrogen, H-H, in contrast to a heteronuclear diatomic molecule like H-Cl. In the first case, rotation of the molecule through $\pi$ radians (180°) produces exactly the same rotational configuration that we started out with, H-H, while in the second case, a new rotational configuration, Cl-H, is produced. (The number of purely rotational symmetry transformations belonging to the point group of H-H is 2.) The rotational partition function in this case is divided by 2:

$$q'_{rot} = \frac{q_{rot}}{2} = \frac{T}{2\theta_{rot}}. \qquad (9.8.1)$$

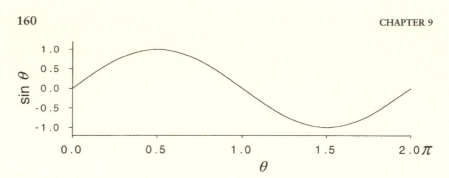

Figure 9.8.1. The rotational wave function for $J = 1$. Starting at any arbitrary point on the wave (except zero) and moving $\pi$ to the left or right produces a change in sign.

Quantum mechanically, the existence of symmetry numbers is a manifestation of the principle that the total wave function must be symmetric or antisymmetric according to whether it governs a boson or fermion. The total wave function can be written as the product of wave functions for all quantized motions within the molecule. The partial wave functions that interest us here are those for rotation and nuclear spin, $\psi_{rot}$ and $\psi_{nuc}$ respectively. For our purposes, we can just think about the product

$$\Psi = \psi_{rot} \times \psi_{nuc}. \tag{9.8.2}$$

In ground molecular hydrogen, the nuclear spins are antisymmetric, one $m_s = -1/2$ and the other $m_s = -1/2$, which means that exchange of the two nuclei, which is equivalent to a rotation of $\pi$ radians, causes $\psi_{rot}$ to change sign. The product $\psi_{rot} \times \psi_{nuc}$ of an antisymmetric function and a symmetric function is antisymmetric (like the product of an odd and an even number). Because the product $\Psi = \psi_{rot} \times \psi_{nuc}$ must be antisymmetric and $\psi_{nuc}$ is antisymmetric, $\psi_{rot}$ must be symmetric. This places restrictions on the rotational quantum number $J$. Only even values of $J$ lead to rotational states that are symmetric, while odd $J$ values lead to antisymmetric rotational states, which are forbidden because they give a symmetric product in equation 9.8.2 (anti $\times$ anti = symm). To see why only even $J$ values are allowed, consider the lowest few rotational wave functions. For $J = 0$, the molecule is not rotating. This is a trivial case. At $J = 1$, the wave function takes the form of the sine wave in figure 9.8.1. The wave function is *antisymmetric* to a rotation of $\pi$ radians. This rotation is forbidden by the symmetry rule. If $J = 2$ as in figure 9.8.2, however, the wave function is symmetric for rotation by $\pi$ radians and the quantum state is allowed. By extension, $J = 0, 2, 4, 6, \ldots$ are allowed and $J = 1, 3, 5, \ldots$ are forbidden. Only half the quantum states available for a heteronuclear molecule are available to H-H because of the forbidden odd

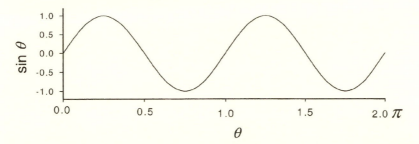

Figure 9.8.2. The rotational wave function for $J = 2$. Starting at any arbitrary point on the wave and moving $\pi$ to the left or right does not produce a change in sign.

$J$ values; hence the partition coefficient, $q_{rot}$, which measures the number of available states, is divided by 2 for H-H. Half the rotational states are lost, which leads to the symmetry number $\sigma = 2$ in the denominator of $q_{rot} = T/2\theta_{rot}$. One might expect lines to be missing from the rotation spectrum of H-H, and indeed they are for truly ground state H-H, called *para*-hydrogen. Only half the usual number of lines are found, those for transitions between allowed states. Forbidden states produce no spectral lines.

Many molecules have symmetries. Ammonia, $NH_3$, has threefold symmetry (for rotation of $\frac{1}{3}(2\pi)$, methane has 12 (three looking down each of four C-H bonds), and so on. In general,

$$q_{rot} = T/\sigma\theta_{rot}, \tag{9.8.3}$$

where $\sigma$ is the symmetry number, 2, 3, or 12 in the examples cited. How does this influence the thermodynamic properties of matter? With only $1/\sigma$ of its rotational states, the partition function $q_{rot}$ (sum over states) is diminished by $1/\sigma$ and any thermodynamic function involving $q_{rot}$ is also diminished.

## 9.9 Bosons, Fermions, and Triplets

The situation is not always as simple as described above, although the principles do not change. Not all molecular hydrogen is *para*-hydrogen. An excited nuclear spin state with a multiplicity of 3 (a *triplet*) exists in which nuclear spins are parallel. It is called *ortho*-hydrogen. Ortho-hydrogen, with parallel spin states, has $\psi_{nuc}$ that is symmetric; hence $\psi_{rot}$ must be antisymmetric. Only odd values of $J$ are allowed. The rotational spectrum of *ortho*-hydrogen has alternate lines missing just as *para*-hydrogen does, but they are different lines. The lines missing in *para*-hydrogen appear in *ortho*-hydrogen and the lines found in *para*-hydrogen are missing in the *ortho* nuclear spin state.

So-called *normal* hydrogen is a 3 : 1 (triplet : singlet) mixture of *ortho*- and *para*-hydrogen which is just what you would expect from the discussion on fine structure in section 7.13 and following). The rotational spectrum of normal hydrogen consists of *ortho*- and *para*-hydrogen spectra superimposed. Lines due to *ortho*-and *para*-hydrogen alternate with intensities in the ratio of 3 : 1, where the intensity due to *ortho*-hydrogen is three quarters of the total and that of the *para*-hydrogen is 1/4 of the total.

Averaging over many quantum states,

$$q_{\rm rot} = \frac{\frac{3}{4}\sum_{\text{odd }J}(2J+1)e^{-\beta hcBJ(J+1)} + \frac{1}{4}\sum_{\text{even }J}(2J+1)e^{-\beta hcBJ(J+1)}}{2}$$

$$= \frac{1}{2}\sum_{\text{all }J}(2J+1)e^{-\beta hcBJ(J+1)};\tag{9.9.1}$$

hence the symmetry number is 2 for normal hydrogen. The common spectroscopic notation $B = k_B\theta_{\rm rot}/hc$ is used to simplify the exponent.

## 9.10 The Einstein Coefficients

There are three ways boson populations can be distributed over two energy levels as in figure 9.10.1.

    1. If a boson population $n_1$ at energy $\varepsilon_1$ in figure 9.10.1 is determined solely by the ambient thermal energy through the Boltzmann distribution, the system is at equilibrium and radiation falling upon it is absorbed in a normal way, causing promotion of bosons from the lower level to the upper level followed by thermal return of energy to the environment. This is ordinary absorption of light of the kind that we use in spectrophotometric analysis.

    2. Because two or more bosons can occupy an energy state, the distribution can show Bose-Einstein *enhancement* of the number of photons in the upper state. The upper state can emit photons *spontaneously*. If, through enhancement, the populations of the two levels are such that the system emits a photon for every photon it absorbs, it is transparent. (Systems may also be transparent if they have no natural frequencies that resonate with the incident radiation.)

    3. If $n_1 > n_0$, the population is *inverted* and the system may *lase*. Lasing involves the *stimulated* emission of one or more photons brought about by an incident photon.

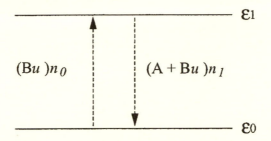

Figure 9.10.1. The Einstein coefficients A and B.

At equilibrium, opposing rates are equal just as they are in a simple chemical reaction

$$\frac{dn_1}{dt} = -(A + Bu)n_1 + Bun_0 \qquad (9.10.1)$$

and

$$\frac{dn_0}{dt} = (A + Bu)n_1 - Bun_0 \qquad (9.10.2)$$

where $u = u(\omega, T)$ = energy density, A is the rate constant (probability) for spontaneous emission, and B is the rate constant for *stimulated emission* through Bose enhancement.

Because equilibrium holds, rate $\uparrow$ = rate $\downarrow$ and

$$\frac{d(n_0 + n_1)}{dt} = 0, \qquad (9.10.3)$$

$$\frac{dn_0}{dt} = \frac{dn_1}{dt} = 0, \qquad (9.10.4)$$

and

$$(A + Bu)n_1 = Bun_0. \qquad (9.10.5)$$

But we also have

$$\frac{n_1}{n_0} = e^{-\beta\hbar\omega}, \qquad (9.10.6)$$

so

$$u(\omega, T) = \frac{A}{B} \frac{1}{e^{\beta\hbar\omega} - 1} \qquad (9.10.7)$$

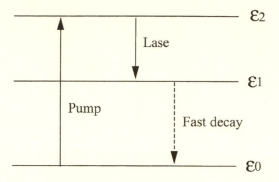

Figure 9.11.1. One arrangement for optical pumping of a laser.

for bosons. From the Planck distribution

$$u(\omega, T) = \frac{\hbar\omega^3}{\pi^2 c^3} \frac{1}{e^{\beta\hbar\omega} - 1},$$

it follows that

$$\frac{A}{B} = \frac{\hbar\omega^3}{\pi^2 c^3} = \frac{8\pi h\nu^3}{c^3}. \tag{9.10.8}$$

## 9.11 Lasers

The most noteworthy practical application of Bose-Einstein statistics is the laser (*light amplification by stimulated emission of radiation*). At high radiation densities $B \gg A$,

$$Bun_1 = Bun_0. \tag{9.11.1}$$

Radiation on the system brings about

$$1 \to 0$$

transitions. If $n_1 \gg n_0$ (population inversion), more stimulated photons are produced than incident photons; hence photon intensity is enhanced. Coherent photons (those with the same direction and phase as the incoming photon) are produced.

A critical issue is, of course, how to achieve the Bose-Einstein enhancement in the first place. An efficient arrangement is shown in figure 9.11.1. An outside source of intense radiation drives the system to a level $\varepsilon_2$ which lases to a level $\varepsilon_1$. Because it is not populated by the exciting radiation,

$\varepsilon_1$ is essentially empty, increasing the probability that the system will lase from $\varepsilon_2$ to $\varepsilon_1$. In an efficient system, $\varepsilon_1$ decays rapidly, giving out thermal radiation. This decay being fast, $\varepsilon_1$ remains essentially unpopulated and the lasing action is not opposed. The initial enhancement process to $\varepsilon_2$ is called *pumping*. The ruby laser ($Al_2O_3$ doped with minute amounts of Cr) can be pumped using a xenon flashlamp.

## 9.12 The Bose-Einstein Condensation

The Heisenberg uncertainty principle tells us that the more accurately we know the momentum $p$ of a particle, the less accurately we can specify its position. The de Broglie wavelength $\lambda_{dB}$ tells us that the proportionality constant between the particle wavelength and the inverse of its momentum is approximately Planck's constant:

$$\lambda_{dB} \cong \frac{h}{p}. \qquad (9.12.1)$$

Schrödinger reasoned that, if material particles have a wavelength, they must be governed by a wave function $\Psi$. Finally, Born's interpretation of the wave function of a material particle tells us that $\Psi^2$ is the *probability* function governing the likelihood of guessing where the particle is, even though we are not able to calculate or observe its exact location.

Putting all these ideas together, Einstein pointed out that as the temperature of a nonrelativistic system approaches 0 K, motion slows down and momentum, in the denominator of equation 9.12.1, approaches zero. This being the case, there must be some temperature low enough that the probability wave of material particles becomes large. If the probability wave is large enough, two things can happen:

1. bosons (atoms) may group together under the same wave function and become a "superatom" and
2. quantum phenomena may become observable in macroscopic behavior.

In a *Bose-Einstein condensate*, we are able to observe quantum phenomena in macroscopic systems for the first time, and to explain otherwise mysterious phenomena such as superfluidity and superconductivity.

## 9.13 The Bose-Einstein Condensation of Metal Vapor (Nobel Prize, 2001)

Bose-Einstein condensing a sample of gaseous atoms entails daunting technical problems solved individually by many groups and collectively by

Cornell and Wieman in 1995. First, gaseous atoms must be confined in some kind of "container." Cornell and Wieman confined rubidium atoms by trapping them in a magnetic field. Magnetic dipoles, the rubidium atoms in this case, can be confined and manipulated by changing the strength and shape of the field. The magnetic field due to more than one permanent or electromagnet is a superposition of them all, permitting its shape and strength to be arbitrarily controlled.

Aside from *magnetic trapping* of atomic dipoles in a field of suitable shape, atoms are confined by *optical trapping and cooling.* Lasers focused on the center of the field are tuned so as to transmit kinetic energy to atoms moving toward the source of the laser beam. Atoms moving away from the laser beam have a different kinetic energy relative to it (Doppler effect) and do not receive a "hit" from the laser. Six lasers normal to the faces of a cube point toward the center of the sample and, averaged over many collisions, they tend to "herd" atoms toward the center of the field.

In addition to helping confine the sample of gaseous atoms to the center of the trap, the laser beams cool atoms by slowing them down. If a laser beam is tuned so as to deliver photons at an energy slightly lower than the resonant frequency of a collection of magnetically trapped gaseous atoms, most of the atoms will not absorb any photons. An atom moving toward the laser source will, however, "see" light at a frequency higher than an atom moving in any other direction by the Doppler shift of the oncoming laser beam brought about by the atom's motion. Provided that the laser is tuned to the right frequency to achieve a match of its own Doppler-shifted frequency with that of the oncoming atoms, fast atoms will be selected for absorption from among the randomly moving mass of atoms. Excited atoms then lose photons at their resonant frequency. The result is that photons are going into the system at slightly lower energy than those coming out, with the net result that the system is cooled.

Atoms are also lost from the field. This leakage is beneficial because the atoms lost are fast-moving atoms from the high end of the energy distribution. Loss of high-energy atoms reduces the average energy of the sample, which brings about further cooling. The result of these cooling methods operating in concert is that the sample can be brought to about $10^{-7}$ K. This temperature is almost unimaginably low. On a centigrade thermometer stretching from Los Angeles to New York, the temperature of a Bose-Einstein condensate would read about 0.7 mm.

When the temperature is lowered to near 0 K, metal atoms do not disappear as photons do; if they are bosons, they crowd (condense) into the lowest-energy state available. Their essential quality is that they are all governed by the same wave function. Formerly incoherent atoms have

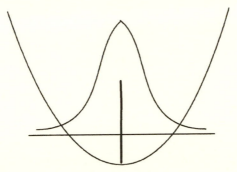

Figure 9.14.1. Ballistically expanded sample containing a BEC. The solid vertical represents what the expansion would look like if the condensate obeyed classical mechanics. It is not observed. The parabola is a schematic representation of the potential energy of a magnetic trap.

become *coherent* in space and in wave function. They have formed a *superatom*. There does not seem to be an upper limit on how many atoms can be Bose-Einstein condensed. Cornell and Wieman condensed about 2000 rubidium atoms.

## 9.14 Ballistic Expansion

One of the difficulties in Bose-Einstein condensate (BEC) production is knowing when you have one. The vapor samples obtained up to now are too small to permit optical analysis. The solution to this problem is analysis of the product of a *ballistic expansion*. In a ballistic expansion, the restraining magnetic field is suddenly released so that the whole sample, Bose condensed atoms and atoms that are not Bose condensed, expands into an essentially free volume. As this process takes place, the BEC is destroyed but it leaves a residuum of low-temperature, slowly moving atoms. While it is not observed directly, the BEC leaves evidence that it was present in the original sample before expansion.

Laser beams passing through a vapor are deflected according to the density of the vapor. Hence a laser scan of the ballistically expanded vapor should show two regions, a disperse region containing fast-moving atoms that were not Bose condensed and a region of high density at the center of the atom cloud comprised of atoms that were formerly Bose condensed, but are now merely slow-moving members of a normal atom collection (figure 9.14.1).

## 9.15 Macroscopic Quantum Effects

Up to now, we have believed in the truth of several quantum mechanical postulates (the existence and properties of the wave function) on circumstantial evidence. Admittedly there is a lot of it, and almost everyone agrees that the standard approach to quantum mechanics is a valid way of describing the submicroscopic world of single atoms and electrons. What of superatoms? Will they be large enough that we shall be able to see (or at least videotape) their macroscopic *quantum* behavior?

Figure 9.14.1 contrasting quantum and classical behavior shows that the answer is yes. In a collection of particles obeying classical mechanics, all motion would cease or become imperceptibly small as the sample arrives at a temperature imperceptibly greater than zero. This state of affairs is represented by the solid probability vertical in figure 9.14.1. All particles would be found precisely at the bottom of the parabolic potential energy well (uncertainty in position = 0) and all would have the same zero energy because they are not moving (uncertainty in momentum = 0). The quantum mechanical predictions are very different. Quantum mechanics says that the system will not go to the bottom of the potential energy well but will retain some zero-point energy and that the probability distribution will not be a vertical line but will be a bell-shaped distribution resembling the square of a Gaussian function like the one in the center of figure 9.14.1. Experimental videotapes of a ballistically expanded BEC by Cornell and Wieman show unequivocally that the *quantum distribution* governs Bose-Einstein-condensed samples of rubidium atoms.

## 9.16 Superfluidity

At about 2.2 K something very remarkable happens to liquid $^4$He. Its physical properties are altered such that its viscosity becomes negligible, causing it to flow readily through porous plugs, "creep" spontaneously out of containers, and support (metastable) vortices. Once again, the heat capacity is indicative of something unusual. Heat capacity as a function of temperature experiences a discontinuity in a region where the curve is shaped vaguely like the lower-case Greek letter lambda in figure 9.16.1. A form of liquid helium called helium-II is observed below the lambda point. It is said to be a *superfluid*. Two classes of rotation experiments exist, one of which can be explained by simple superfluidity while the other requires the notion of topological winding. We shall consider an example of each.

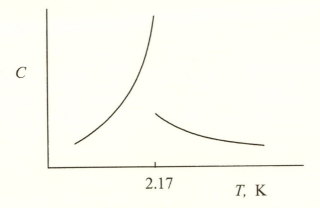

Figure 9.16.1. The heat capacity of helium-II as a function of temperature near 0 K. The curve is said to have a lambda point.

### Experiment 1.

Take a small cylindrical cup of helium at 3 K, where it is an ordinary liquid, and cause it to rotate about its vertical axis at a small angular speed $\dot{\omega}$. Viscous drag at the interface between the cup and the outermost liquid layer causes the liquid to be accelerated to $\dot{\omega}$. After a while, the cup and the liquid are rotating at the same speed. Now carry out the same experiment well below 2.17 K where helium has largely Bose condensed into superfluid helium-II. The He-II refuses to rotate with the container even if rotation of the cup were to be continued forever. This is a class 1 experiment because we can explain it simply by saying that the viscous drag at the interface between liquid and container is absent in He-II. No viscous drag, no angular acceleration. The liquid never rotates with its container.

The obvious question in experiment 1 is why viscous drag on He-II should drop to zero, and the answer follows directly from what we have already found to be true of Bose-Einstein condensates. All the particles at the liquid-container interface are governed by the same wave function. You cannot increase or decrease their angular speed one at a time as you can with a normal liquid. They have to be jolted into or out of their rotational state all at once. This requires an energy input far in excess of the meager thermal energy available (don't forget, we are near 0 K). If we supply energy by warming the system, we lose the Bose-Einstein condensate.

*Experiment 2.*

In one of a second category of experiments, we choose different rotational speeds of the container prior to cooling the system below 2.17 K. As we vary the angular speed, we come upon critical speeds $n\dot{\omega}_\lambda$ at which the superfluid He-II rotates in equilibrium with its container even very near 0 K, as a normal liquid would. The situation is like that of the electron in a hydrogen atom which (under the old Bohr theory) can have one angular speed (orbit) or another angular speed, but not any speed in between. This is *quantized rotation of a macroscopic sample.* At very low speeds, the condensate will not rotate at all as in experiment 1. It is in the rotational ground state $\dot{\omega} = 0$. At higher cup speeds, the condensate rotates at $\dot{\omega}_\lambda$, $2\dot{\omega}_\lambda, 3\dot{\omega}_\lambda, \ldots$, but not at $0.87\dot{\omega}_\lambda$, $1.5\dot{\omega}_\lambda$, or any other nonintegral multiple of $\dot{\omega}_\lambda$. In a startling variation of this experiment, if the container rotates at a speed slightly more than $\dot{\omega}_\lambda/2$, He-II begins to circulate, not at the cup speed of $\dot{\omega}_\lambda/2$ but at $\dot{\omega}_\lambda$. In other words, He-II is circulating at *twice the angular speed of the rotating container.* That takes some explaining. Zero viscosity will not do.

## 9.17  Order Parameters

Order parameters, as the name implies, are a measure of internal order in a mechanical system. For example, a hypothetical ferromagnetic solid having all electron spins oriented in the same direction would have an order number of 1.0. Order parameters do not have to be simple integers. They can be vectors in 2- or 3-space for surfaces or solids, or they can be complex for rotation. A rotational order parameter has a wave function $\Omega = e^{-i\omega}$ associated with it that must be continuous. The variable $\omega$ is the angle of rotation. A stable wave function can be depicted as a helix wound about the circumferential path of rotation in an integral number of turns; hence the origin of the term *winding number.* There is a stable rotational state at $\Omega = 0$, one at $\Omega = \pm 2\pi$, one at $\Omega = \pm 4\pi$, and so on, but not at $\Omega = 0.87\pi$, $\Omega = \pm 1.5\pi$, or other noninteger multiples. If the winding number is not an integral multiple of $2\pi$, $\Omega$ suffers a discontinuity. Winding numbers are quantum numbers for rotation, $n = 0, \pm 1, \pm 2, \ldots$, where $\pm$ denote opposite directions of rotation.

Electrons in atomic hydrogen moving from one Bohr orbit to another at room temperature (to choose a simple model) absorb photons one at a time. Changing a macroscopic sample of a Bose condensate by one rotational quantum state, however, would require that all particles change state at once, taking up $\sim 10^{23}$ times as much energy. The activation energy is too high. The rotating superfluid is locked in a metastable rotational state from which it cannot escape. To change a rotational state, warm the superfluid, accelerate or decelerate it, then cool it down again.

A subject of speculation has been whether Bose-Einstein condensates of atomic gases will be superfluids at an appropriate temperature. Recent experiments on gaseous rubidium Rb show that they are. Recall that both gases and liquids are fluids (they flow); hence Bose-Einstein-condensed atomic vapors are *supergases*.

## 9.18 Superconductivity

The history of superconductivity goes back almost as far as the history of quantum theory itself. In 1908 Kamerlingh Onnes, wishing to study the heat capacities of metals at low temperatures, chose mercury, which can be obtained at very high purity by distillation. He soon found that at a critical temperature $T_c \cong 4.1\,\mathrm{K}$, slightly below the boiling point of helium, mercury shows *zero resistance to electrical current flow*. Subcritical mercury is said to be a *superconductor*.

The mechanism of ordinary conduction by metals involves passage of electrons through a lattice of positive ions and the core electrons associated with them. The positive metal ions vibrate about their lattice sites, and the overall flow of electrons is impeded by collisions between freely flowing conductivity electrons and the ion-electron core. Not many of these collisions take place, because the heavy ion lattice vibrates much more slowly than the speed of a passing electron, but there are enough collisions to resist electron flow according to Ohm's law $E = IR$, where $E$ is the potential across the conductor, $I$ is the current or electron flow, and $R$ is the resistance of the conducting metal. Metals are good conductors but they are not perfect.

Below $T_c$, a new mechanism or "path" for conduction becomes available to electrons passing through the conductor. At this low temperature, thermal agitation becomes so small that spin attraction overcomes charge repulsion in electrons, and opposite-spin electrons pair up to form what are called *Cooper pairs*. Distortion of the positive-ion lattice causes a slight increase of positive charge surrounding each electron. This charge shields interelectronic repulsion and stabilizes pair formation. Cooper pairs, having spin $s = +\frac{1}{2} - \frac{1}{2} = 0$ are bosons even though the electrons that comprise them are fermions ($s = \pm\frac{1}{2}$). Because they are bosons, Cooper pairs can form a Bose-Einstein condensate. Now for lattice vibrations (low-energy, long-wavelength phonons) to impede current flow due to Cooper pairs, they would have to drive the entire condensate of Cooper pairs back against the current flow. This does not happen. There is no resistance to a flow of Bose-condensed Cooper pairs, and the electrical resistance of the metal, $R$, drops to zero.

By analogy with catalysis in chemical reactions, when a low-energy reaction path is opened, that is the path the reaction will take. When a zero-resistance conduction mechanism for current flow through a metal is opened via Cooper pair condensates, that is the mechanism electrons will take, even at temperatures above 0 K. The appearance of superconductivity is not gradual as the temperature is lowered from $T_c$ toward 0 K. Rather, it is sharp and appears as the new conduction path is opened at $T_c$. For obvious practical reasons, the quest for high-temperature superconductors is a very active research field.

## 9.19 Stopped Light

A sign over a biker bar in Northern California advises its patrons "186,000 miles per second; IT'S THE LAW." But is it really? Newton believed that dispersion of white light into its component colors by a prism depends on the reduced speed of light in glass, which, in turn, depends on the color of the light. Today, we agree with the essentials of Newton's ideas on dispersion. Without wishing to denigrate the public spiritedness of the proprietors of our favorite biker bar, we should correct the sign to read "186,000 miles per second in a vacuum; IT'S THE LAW."

The speed of light in water or in glass is less than $c$ by a factor of about 1.5. Until recently, attempts to slow light much more than this have run into the difficulty that the increased refractive index necessary for slow light is accompanied by an increased light absorption by the optical medium. In 1999, however, Hau and coworkers slowed light to $v = 17 \, \mathrm{m \, s^{-1}}$ (about the speed you can achieve on a bicycle) in a Bose-Einstein condensate by *laser-dressing* Na atoms at temperatures below 435 nK ($435 \times 10^{-9}$ K). In more recent experiments (2000), the same research team *stopped light entirely* in a laser-dressed Na condensate.

Laser dressing involves exciting Na atoms with a laser beam so that emission and absorption of light from atoms precisely cancel, in other words, imposing a laser-induced transparency on the sample. The ground state sample in a magnetic field has a two-level fine structure owing to the energy difference arising from interaction of the field with the electron magnetic moment $\pm$ the nuclear magnetic moment. Imposition of a coupling laser field converts the fine structure to a noncoherent superposition. The noncoherence exists because component states have been adjusted by tuning the coupling fields to be exactly out of phase with one another. They suffer destructive interference. The transition dipole moment from the noncoherent state to the excited state is zero; hence photon absorption does not occur, and the sample is transparent.

At the same time, the refractive index of the sample sharply increases so that the speed of light within it decreases accordingly. Once the sample has been laser dressed, an independent (perpendicular) probe source produces pulses that travel slowly through the condensate. By varying external parameters, including the intensity of the probe pulse, photons can be slowed arbitrarily or even stopped for about a millisecond ($\sim$1 ms), then released at normal photon speed. Information imprinted on the light pulse, by amplitude modulation, for example, is retained by stopped light in the Bose condensate and is transmitted later by the emergent beam of photons. Information retention by stopped light may have applications in data storage. Although $\sim$1 ms may not seem like a very long time to store data, most computers can carry out many millions of operations in that amount of time.

## 9.20 Vortices

If you stir water or, better yet, iced tea vigorously (but not too vigorously), you may see tiny "swirls" parallel to the axis of stirring but not coaxial with it. (They are more visible in tea.) These are vortices. A vortex is a metastable state intermediate between smooth fluid rotation and turbulence. All fluids, air included, display vortex formation. A tornado is a vortex. A subject of speculation has been whether Bose-Einstein (B-E) condensates will have metastable vortices. Recent experiments show that they do. Quantum rules govern vortices in B-E condensates.

Calculating the moment of inertia of a cylinder is a routine problem in calculus. So, in principle, is measuring the moment of inertia $I = mr^2$ of a cylindrical cup of fluid by suspending it from a torsion wire and observing its resistance to angular acceleration. For simplicity, let us think about an annular shell (constant $r$) of liquid He. At some temperature between 0 K and $T_\lambda$, the cup behaves as though it were filled with two liquids, superfluid He-II and normal liquid He. Its moment of inertia under low acceleration will be less than that of the same ring filled with normal He because superfluid He-II does not rotate. He-II is essentially "not there." The ratio $I/I' > 1.0$ where $I$ is the moment of inertia of the normal fluid and $I'$ is the moment of inertia of the mixture. Because $I \propto m$ and the density $\rho \propto m$ for a constant-volume shell,

$$I/I' = \rho/\rho' \qquad (9.20.1)$$

where $\rho$ is the mass density of the normal fluid and $\rho'$ is the inertial density of the mixture. If the fluid is normal, there is no superfluid present and

$\rho = \rho'$, but in a mixture, the mass of normal fluid is reduced by the amount of normal fluid that has been converted to superfluid. Its density is diminished accordingly ($\rho' < \rho$). The *superfluid density* $\rho_s$ is defined as the difference between the pure normal fluid density and the inertial density of the mixture, $\rho_s \equiv \rho - \rho'$, that is, the density that has been given over to the superfluid. The assumption of additivity, that the mass density is the sum of the inertial density due to normal He-II plus the superfluid density, $\rho = \rho' + \rho_s$, has been made in this model.

The *momentum density* for a normal fluid is $g = \rho v$ but for a superfluid it is

$$g = \rho' v' + \rho_s v_s, \tag{9.20.2}$$

which is the origin of the term *superfluid velocity* $v_s$. The circulation of a Bose-Einstein condensed vortex is quantized according to the cyclic integral of its superfluid velocity over the radius $r$ of any *vortex contour*

$$K = \oint v_s d\,r = \frac{\hbar}{m} 2\pi n = \frac{h}{m}\,n, \qquad n = 0, 1, 2, \ldots, \tag{9.20.3}$$

where $v_s$ is the superfluid velocity about the closed contour. The order number for circulation of a vortex is an integral multiple of $2\pi$. A vortex line must close on itself to form a vortex loop without a discontinuity in the wave function for fluid rotation.

One naturally wonders how a sample about the size of a small grain of pepper (small but still qualifying as "macroscopic") is stirred. One way is to impose an asymmetric alternating magnetic field on the condensate with its antinodes at some distance from the vertical axis. The field is then rotated about the vertical axis in much the same way as one would stir a glass of iced tea. In this way multiple vortices have been produced (see cover illustration).

Photons in the vicinity of a vortex will be swept along with it like a henhouse roof in a tornado. This process is somewhat analogous to matter being swept into a black hole. One's ability to create and control tiny vortices in Bose-Einstein condensates presents an intriguing model of real black holes, with the possibility of studying the physics of black holes in their miniature laboratory models. According to some recent theory, black holes come in all sizes and may be as small as elementary particles. Indeed, they may *be* elementary particles (Greene 1999).

## PROBLEMS

**9.1.** Show that Planck's equation reduces to the Rayleigh-Jeans equation at low frequencies.

**9.2.** Show that Planck's equation reduces to the Wien equation at high frequencies.

**9.3.** In how many ways can the letters in the name "einstein" be arranged?

**9.4.** Place $N$ photons into $C$ cells with no constraints on the number of photons that can go into a cell. Separate the filled cells into groups: $G(0)$ containing no photons, $G(1)$ containing one photon, $G(2)$ containing two, and so on.
(a) How many ways can this be done?
(b) Give a logical argument showing that your answer to (a) is correct.

**9.5.** Carry out the steps in going from equation 9.7.4 to 9.7.5.

**9.6.** The function $y = x$ is antisymmetric over the interval $[-1, 1]$ and the function $z = x^2$ is symmetric over the same interval (show this). Show that the function $\Psi = z(x)y(x)$ (a symmetric function times an antisymmetric function) is antisymmetric. Try this for other combinations of functions. Does the rule "symmetric times antisymmetric" hold for the functions you have chosen?

**9.7.** Leave $Bu$ out of the term $(A + Bu)n_1$ in equation 9.10.5, that is, ignore stimulated emission. Complete the derivation of an equation corresponding to equation 9.10.8 without this term. What do you get?

**9.8.** What are the units of the Einstein coefficients A and B?

**9.9.** Show that equation 9.10.7, and hence the ratio of Einstein coefficients, follows from equation 9.10.6.

**9.10.** One of the most intriguing characteristics of superfluid He-II is that it spontaneously "creeps" into or out of containers according to the relative liquid level inside and outside the container (figure 9.1). Give a qualitative argument as to why superfluids creep.

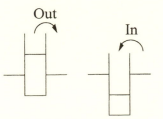

Figure 9.1. Superfluid creep.

# Bibliography

Barrante, J. R. 1998. *Applied Mathematics for Physical Chemistry*. 2nd ed. Prentice-Hall, Englewood Cliffs, NJ.

Bose, S. N. 1924. Plancks Gesetz und Lichtquantenhypothese, *Z. Phys.* 26, 178–181.

Einstein, A. 1905. Über einen die Erzeugung und Verwandlung des Lichtes betreffenden heuristichen Gesichtpunkt, *Annalen der Physik* 17, 132–166.

——— 1907. Die Planckshe Theorie der Strahlung und die Theorie des specifichen Wärme, *Annalen der Physik* 22, 180–190.

Feynman, R. P., Leighton, R. B., and Sands, M. 1965. The Feynman Lectures on Physics, vol. III. Addison-Wesley, Reading, MA.

Greene, B. 1999. *The Elegant Universe*, Random House, New York.

Huang, K. 2001. *An Introduction to Statistical Physics*. Taylor and Francis, New York.

Jammer, M. 1966. *The Conceptual Development of Quantum Mechanics*. McGraw-Hill, New York; 2nd ed. 1989; Tomash Publishers / American Inst. of Physics, New York.

Kittel, C., and Kroemer, H. 1980, rev. ed. 2003. *Thermal Physics*. Freeman, New York.

Kondepudi, D. K., and Prigogine, I. 1998. *Modern Thermodynamics: From Heat Engines to Dissipative Structures*. Wiley, New York.

Kuhn, T. S. 1987. *Black-Body Theory and the Quantum Discontinuity 1894–1912*. University of Chicago Press, Chicago, IL. Many early references can be found in this well-documented work.

Lummer, O., and Pringsheim, E. 1899. Die Vertheilung der Energie im Spectrum des schwartzen Körpers und des blancken Platins, *Verhandl. Deut. Physik. Ges.* 1, 215–235. (See also Richtmeyer et al. 1955.)

Mather, J. C., Cheng, E. S., Eplec, R. E., Jr., Isaacman, R. B., Meyer, S. S., and Shafer, R. A. 1990. *Astrophysical Journal* 354 L37–L40.

McQuarrie, D. A. 1983. Quantum Chemistry. University Science Books, Sausalito, CA.

Rayleigh, see Strutt, J. W.

Richtmeyer, P. R., Kennard, E. H., and Lauridsen, T. 1955. *Introduction to Modern Physics*. 5th ed. McGraw-Hill, New York.

Strutt, J. W. (Lord Rayleigh), 1905, *Nature* 72; 54–55. There is no doubt that Rayleigh was aware of this problem long before he published his paper specifically citing the blackbody radiator. He had carried out an analogous derivation on standing sound waves as early as 1877 (Jammer 1989).

## Web

http://aether.lbl.gov  Go to Cobe, go to Introduction, go to CMB Intensity Plot.

# Index _____

absorptivity, 33
amplitude constant, 13
amplitude, 6
arrangement, 110
Avogadro's number, 67, 80, 87

ballistic expansion, 167
barometric equation, 22
BEC, 165
biker bar, 172
blackbody, 30, 33, 41, 43, 149
blackbody energy density, 61
blackbody radiation energy, 54
blackbody spectrum, 49 ff
black hole, 174
bolometer, 44
Boltzmann distribution, 21, 103, 112, 125
Boltzmann equation, 98
Boltzmann, L. E., 1
Boltzmann's constant, 23, 26, 67, 127
Bose, S., 97
Bose-Einstein condensation, 165
Bose-Einstein counting, 117
Bose-Einstein distribution, 100, 103, 120, 148
Bose-Einstein statistics, 98
bosons, 103, 119, 121, 123, 161, 162
boundary conditions, 13

canonical partition function, 128
Carnot, S., 126
Cartesian coordinates, 70
characteristic temperature, 157, 158
charge, electronic, 68
chemical potential, 122, 152, 153
COBE satellite, 46
combinations, 105
configuration, 104, 105, 126
conservative systems, 5, 11

Debye cutoff frequency, 79
Debye equation, 79, 82

Debye integral, 84
Debye temperature, 81, 82, 84, 147
Debye theory, 77
Debye third power law, 90
Debye-Sommerfeld equation, 146
degeneracy, 19, 58
degenerate levels, 103, 112
degrees of freedom, 23, 70, 80
density of photon modes, 150
density of states, 141, 154
destructive interference, 17
diamond, 71
Dirac delta function, 135
distribution, number density, 79
Doppler shift, 166
Dulong and Petit, law of, 1, 28, 70

Einstein, A., 1, 62, 70
Einstein coefficients, 162, 163
Einstein equation, 102
Einstein frequency, 73
Einstein temperature, 73
Einstein theory, difficulties, 75
electromagnetic radiation, 30, 149; speed of, 10
electromagnetic spectrum, 32, 41, 43
electron gas, 132, 144, 145
electronic energy (Fermi gas), 140. *See also* Fermi gas
electronic heat capacity, 134, 136
emissivity, 36
energy density, 30, 35, 56
energy ladder, 99
enhancement, 162
enthalpy, 91
entropy, 37, 90, 122, 126, 129, 153

faraday, 68
Fermi-Dirac counting, 113
Fermi-Dirac distribution, 100, 103, 116, 132, 135, 145
Fermi gas, 137–138, 144

Fermi sea, 132, 134, 144, 145
Fermi temperature, 133
fermions, 103, 113, 121, 123, 141, 161
Feynman, R. P., 149
filled energy states, 143
fine structure, 110
flux, 34
forbidden quantum state, 160
forbidden wave, 78
frequency, 10, 56
fundamental (thermodynamic) equation, 129
fundamental frequency, 15
fundamental mode, 7

Gibbs chemical potential, 91
Gibbs, J. W., 153

harmonic, 14
harmonic oscillator, 2, 4 ff, 32, 64, 70, 98, 149, 156
heat capacity, 1, 28, 70, 72, 74, 90, 101, 135, 145, 148
He-II, 169
helium, 169
Hooke's law, 70, 73

intensity, 34
interference, 17
isotropic harmonic oscillator, 70

kinetic energy, 24
kinetic energy density, 37
Kirchhoff, G., 30
Kirchhoff's law, 33

Lagrange's method, 108, 116, 120, 151
lambda point, 169
Langley, S. P., 41
laser, 164
laser dressing, 172
lasing, 162
lattice vibrations, 146
linear homogeneous partial differential equation, 12
Lummer and Pringsheim, measurements by, 46, 49

magnetic trapping, 166
Millikan, R. A., 95
mode, 7, 56

modes, 63
molar heat capacity, 1, 27
molar volume, 77, 87
moment of inertia, 156
monochromatic radiation, 32
most probable configuration, 104

Newton, I., 4
Newton-Hooke equation, 5, 12
normal mode, 16
nuclear spin, 160
number density, 23, 56, 59, 66, 77, 80, 87

occupied orbitals, 142
optical pumping, 164
optical trapping, 166
orbital, 138, 139. *See also* occupied orbitals
order parameters, 170
ortho hydrogen, 161
oscillator, virtual, 62
oscillatory modes, 57
overtone, 14

*para* hydrogen, 161
particle in a box, 133, 136
particle partition function, 99
particles, 149
partition function, 99, 128, 157
Paschen, F., 45, 49, 53
permutations, 104
phonon density, 154
phonons, 154
photoelectric effect, 94
photon gas, 96
photon modes, 150
photon states, 150
photons, 94, 149
Planck, M., 30
Planck distribution, 155
Planck equation, 47, 62, 66, 152
Planck's constant, 67, 96
potential energy, 5

quanta, 62
quantum effects, 168
quantum harmonic oscillator, 98
quantum number space, 137, 150
quantum statistics, 94, 98

radial frequency, 10
radiant excitance, 40

radiation density, 30, 32, 39, 53
radiation intensity, 34
Rayleigh-Jeans equation, 60, 87
root mean square, 25
rotational orientations, 159
rotational partition function, 158
rotational spectroscopy, 155
rotational wave function, 160

Schroedinger equation, 136
separation constant, 12
separation of variables, 11
simple harmonic oscillator, 4 ff
Simpson's rule, 85
sound velocity: longitudinal, 88; transverse,
    88
sound, speed of, 78, 86, 87
specific heat, 1
spectrum, 15, 79
spectrum, electromagnetic, 32, 41, 43
spin, 141, 160
standard free energy change, 92
state, 56, 143
state function, 37
statistical distribution, 98
statistical weight, 106, 119
Stefan-Boltzmann law, 36, 39
Stefan's law, 36, 53
stimulated emission, 163
Stirling's approximation, 107
stopped light, 172
stopping potential, 94
superatom, 167
superfluid, 168
superfluid density, 174
superfluid velocity, 174

supergas, 171
superconductivity, 171
superposition, 12, 16
symmetry numbers, 155, 159

tension, 9
thermal radiation, 30
thermodynamics, first law, 153
translational energy, 27
transparency, 172
triplets, 161
trivial solution, 13

uncertainty principle, 165
undetermined multipliers, 108

vibrating string, 7, 10
vibrational characteristic temperature, 157
vibrational energy, 27, 101
vibrational partition function, 157
virtual oscillators, 62
viscosity, 160
vortex contour, 174
vortices, 173

wave equation, 7; general solution, 15; space
    part, 17; time dependent, 15, 19; time
    independent, 12, 18; time part, 17
wave function, 160
wave motion, 9
wave number space, 58
wave numbers, 56
wavelength, 10
waves, 149
Wien, W., 30, 49, 51, 53
winding number, 170

## DATE DUE

| MAR 1 2 2009 | | | |
|---|---|---|---|
| | | | |
| | | | |
| | | | |
| | | | |
| | | | |
| | | | |
| | | | |
| | | | |
| | | | |
| | | | |
| | | | |
| | | | |
| | | | |
| | | | |
| | | | |
| | | | |

Demco, Inc. 38-293